DYNAMIC PROGRAMMING
A Practical Introduction

MATHEMATICS AND ITS APPLICATIONS
Series Editor: G. M. BELL,
Professor of Mathematics, King's College London, University of London

STATISTICS, OPERATIONAL RESEARCH AND COMPUTATIONAL MATHEMATICS Section
Editor: B. W. CONOLLY,
Emeritus Professor of Mathematics (Operational Research), Queen Mary College, University of London

Mathematics and its applications are now awe-inspiring in their scope, variety and depth. Not only is there rapid growth in pure mathematics and its applications to the traditional fields of the physical sciences, engineering and statistics, but new fields of application are emerging in biology, ecology and social organization. The user of mathematics must assimilate subtle new techniques and also learn to handle the great power of the computer efficiently and economically.

The need for clear, concise and authoritative texts is thus greater than ever and our series endeavours to supply this need. It aims to be comprehensive and yet flexible. Works surveying recent research will introduce new areas and up-to-date mathematical methods. Undergraduate texts on established topics will stimulate student interest by including applications relevant at the present day. The series will also include selected volumes of lecture notes which will enable certain important topics to be presented earlier than would otherwise be possible.

In all these ways it is hoped to render a valuable service to those who learn, teach, develop and use mathematics.

Mathematics and its Applications
Series Editor: G. M. BELL,
Professor of Mathematics, King's College London, University of London

Anderson, I. — **Combinatorial Designs: Construction Methods**
Artmann, B. — **Concept of Number: From Quaternions to Monads and Topological Fields**
Arczewski, K. & Pietrucha, J. — **Mathematical Modelling in Discrete Mechanical Systems**
Arczewski, K. and Pietrucha, J. — **Mathematical Modelling in Continuous Mechanical Systems**
Bainov, D.D. & Konstantinov, M. — **The Averaging Method and its Applications**
Baker, A.C. & Porteous, H.L. — **Linear Algebra and Differential Equations**
Balcerzyk, S. & Jösefiak, T. — **Commutative Rings**
Balcerzyk, S. & Jösefiak, T. — **Commutative Noetherian and Krull Rings**
Baldock, G.R. & Bridgeman, T. — **Mathematical Theory of Wave Motion**

Series continued at back of book

DYNAMIC PROGRAMMING
A Practical Introduction

DAVID K. SMITH M.A., Ph.D.
Department of Mathematical Statistics and Operational Research
University of Exeter

ELLIS HORWOOD
NEW YORK LONDON TORONTO SYDNEY TOKYO SINGAPORE

First published in 1991 by
ELLIS HORWOOD LIMITED
Market Cross House, Cooper Street,
Chichester, West Sussex, PO19 1EB, England

A division of
Simon & Schuster International Group
A Paramount Communications Company

Printed and bound in Great Britain
by Robert Hartnolls, Bodmin, Cornwall

British Library Cataloguing in Publication Data

D. K. Smith
Dynamic Programming: A Practical Introduction
CIP catalogue record for this book is available from the British Library
ISBN 0–13–221797–X (Library Edn.)
ISBN 0–13–221805–4 (Student Edn.)

Library of Congress Cataloging-in-Publication Data available

Table of contents

Introduction

This book is intended for students who are prepared to think; it is aimed at undergraduates and graduates on taught courses in operational research and management science who want to explore the ideas of one tool from the armoury of tools that operational research scientists have at their command. Dynamic programming is just that – a tool. It is not, nor can it ever be, a "Black box" in which a computer is used, first, to read in some numbers and then, after a suitable length of time, to produce some further numbers and an answer. Equally, the reader of a book on dynamic programming will not become an expert by memorizing every word, every number and every page between the covers. A little thought will be needed.

Hence the title of the book is "Dynamic Programming: a Practical Introduction". Behind each part of it is the hope that the reader will try and follow up the introductory coverage and practice some aspect of the material in greater depth, to see how problems can be posed in a way that makes the approach of dynamic programming most appropriate. Some readers may recognize ways that the ideas can be made more realistic; others may see how dynamic programming interrelates with other branches of the mathematics associated with operational research – linear and integer programming, network methods, inventory models. The title is appropriate for the author as well. Writing has meant working through material which was familiar from teaching undergraduates and seeing it with fresh thoughts and ideas. It is not very easy, in the limited time that university courses offer, to develop all the ideas that are taught and gain a full practical insight. Spending time looking at dynamic programming ideas has led to the inclusion of comments on the sensitivity of solutions to the data, material which is seldom covered in other books.

In preparing the material I have been grateful to many other people. I learnt about dynamic programming from John Norman's lectures and books, from the textbooks of Nicholas Hastings, Peter Whittle and (above all) the late Richard Bellman. My understanding was stretched through teaching the concepts on several courses at the University of Exeter and in using some examples on external lectures (including those to sixth-formers – 12th grade in the USA). Feedback from the audiences helped in the writing of this book. My especial thanks to those students, Robert Benson, Matthew Scribbins and Julie Whitehouse, who read a draft and made valuable suggestions for improving several sections. Collaboration with others in research has pointed to the value of dynamic programming as a way of looking at problems. Acting as an editor for the International Abstracts in Operations Research has made me aware of the frequency with which dynamic programming provides the tool for investigating industrial and commercial decision making. I have been very grateful for the work of Donald Knuth (1984) in developing TEX, used for the preparation of the body of the book, and for the programmers who produced software that enabled the tables and diagrams to be produced in TEX. Staff from Ellis Horwood have regularly given very helpful advice and assistance. For their support while I was writing, thanks go to my university colleagues – even to the one who announced that "all dynamic programming is concerned with finding the shortest – or longest – path"! And for her continued support, encouragement and prayer, my heartfelt thanks to my wife Tina, my chosen companion on the longest path of all.

Next to any text book on a student's desk, there should be writing materials; the author has assumed that the reader will have these available. Then, mathematical formulations can be checked and exercises worked through. All will be part of the process of education – literally, "leading out". In the words of Gene Woolsey:

"Training is the absorption of skills;
Education is the acquisition of new hungers." †

The author's hope is that many readers will develop ravenous appetites!

Knowledge of some other aspects of operational research is assumed in several places in this book. There are numerous good general texts on the principles and practice of operational research, and it would be hard to recommend one for the reader who needs to fill in a gap in their understanding. A visit to your friendly neighbourhood library should yield a book to fill in the assumptions made about operational research – and it may even produce further examples of dynamic programming!

† in Interfaces **19** (May-June 1989) page 46

1

What is Dynamic Programming?

1.1 Decisions in sequence

What is the best way to travel from where you live to where you work? Or what is the best route to follow when you travel from home to a holiday destination? What is the best itinerary for the vehicle that delivers food from the suppliers to the shop nearest to your home?

Each reader can fill in the details of these questions and each can make some kind of answer to them. There will be two difficulties in many answers. First, the questions have not defined what is meant by "best"? Second, what should be the form of the answer; how much detail is needed?

I live three and a half kilometres from my office. Most days I travel to work by bicycle, normally following the same route. This can be described as a succession of roads and streets to be followed, or as a series of places that my route will pass through. It is "best" because it is the shortest route that avoids the city centre (where the motor traffic is heavy, and drivers sometimes ignore cyclists). Shortest means the least distance, with the condition that the areas of congestion are avoided; it also means that the expected time taken for the journey is least, and (as far as I can judge) it is the safest route for a cyclist. When I travel by car, then I will follow a slightly different route, because my definition of "best" for the car journey is the route which makes the time taken, on average, as small as possible. The route that I take in a car is, I believe, the one which uses the smallest amount of fuel and so has the least average cost. When I walk, I usually take a third route, which is the shortest possible on public paths – and shortest in the time it takes (again, on average).

"Best", for this simple journey, has several different meanings. Least distance; least time; least cost; safest; all could have constraints. For a longer trip, there might be other definitions of "best". Travelling within Europe, a journey could be made up of several stages; a walk to the coach-station; by coach to the airport; flying to another country by plane; by underground train to a railway station at the end of the flight; a train ride; a further walk to complete the adventure. This could be defined as "best" as being the one which is the least expensive, subject to taking less than a day to complete it, and to not having to carry luggage more than a certain distance. For a millionaire, "best" might be the journey which took the least time – taxis and private jets replacing the variety described above. For a holiday-maker, "best" could be the one which allowed visits to interesting tourist attractions on the way. For a lover of good food, "best" might be the route which passed high quality restaurants at mealtimes.

The variety of meanings of a simple word is an important part of the application of operational research in practice. Within an organisation, different people will have a variety of ideas about what is best for their business. It may be to make as much income as possible; it may be to satisfy customers' requests; it may be to minimise the financial costs of a part of the enterprise. Selecting which meaning is intended is an important part of the skills of the operational research scientist's work. So is the task of explaining the solution clearly; for a journey, it will be described as a series of actions, which have to take place in sequence, one after the other. The description of the actions will make clear what is to be done whenever (or wherever) there is a choice, either of roads (on the short journey), or of means of transport (on the longer trip). Whichever happens, the solution to the problems of the first paragraph will be an ordered set of instructions, to be followed in sequence. Dynamic programming is an approach to problems like this, and devising answers to questions of sequential choice is the aim of this book.

The examples of personal experience in travelling may seem to be trivial. Their consequences could not be claimed to have an effect on one's life. But some choices do lead to changes whose sequels do not disappear rapidly. For example, an advertisement for a job may lead to a chain of decisions. First, whether or not to write off for more details and a form on which to make one's application. Then whether or not to apply. Then, possibly, whether or not to attend the interview, and at the end, whether or not to accept the position. And that's being optimistic; what happens when there are several jobs which look interesting and you have to make decisions about each one alongside all the others? Of course, if you have accepted a job, then you are not really free of the series of decisions; you still have to decide how long to stay in the position before starting the whole process again and applying for other posts!

Because we live in a world where time progresses forwards, and our brains are unable to handle many things at the same time, all our personal decisions are made sequentially. We may feel that our minds have several choices open simultaneously, but in practice they have to be dealt with one after the other. Our major personal decisions affect us in many ways: where we live; our way of life; our standard of living. Usually (and fortunately) it is impossible to foresee every consequence of every decision that we may take.

1.2 Sequential decision making in business and commerce

The examples in section 1.1 have been drawn from personal observation and experience. Similar problems can be found readily enough in industry and commerce. There the consequences will be more important. The financial well-being of a business depends on the correct decisions being made at all times. Many such sequential decisions are extremely complex, requiring thorough knowledge of the company and its place within an industry. Other kinds of decision are much more general and so similar sorts of decision problem can be found in many settings. Thus problems of deciding how much of a product to manufacture are certain to appear in many organisations. Problems regarding investments and problems about ordering policies for goods will occur in many sectors of commerce. So we can imagine a problem such as the following without being specific about what product is being described, nor in which industrial context the problem can be found:

> A company has 4 items on hand and will have to manufacture further items to meet demand over the next four weeks. Orders average 5 items per week but are randomly distributed between 3 and 7 with a uniform distribution. Three kinds of costs may be incurred by the company: setup costs, which are the amount it must pay to start a production run in any week; unit costs, per item made; storage (or holding) costs for each item held unsold from one week to the next. These vary from week to week in the way that is shown in Table 1.1.

Table 1.1: Future demand and costs for a product

	Week 1	Week 2	Week 3	Week 4
Setup cost	100	120	100	140
Unit cost	6	6	5	7
Holding cost	1	2	2	1

> What plans should the company make for production over the next four weeks if it wishes to minimise its expected expenditure?

Outlined in such a way, the company has to make weekly decisions, and the state of the company – as measured by its inventory level – depends on the previous week's decision, as well as on the random demand for the product. There is a sequence of decisions, one each week, to be made in a particular order. The cost depends on what decision is taken, and it is affected by the uncontrollable random demands for the items.

In this book we intend to explore ways of solving problems like these; problems where decisions are made one after another, where the sequence is taken over space (as in a journey) or time (as in a commercial problem). Dynamic programming is not limited to situations where it is so obvious that decisions need to be taken as a series. We shall also look at problems where there are several decisions to be made, but which seem to require that these should be taken at the same time. However, we shall see that the approach of dynamic programming can be used to solve some instances of such problems. It is often possible to shape the problem into one where the simultaneous decisions are replaced by a sequence of decisions, to be thought about one at a time, one after another.

1.3 Looking at dynamic programming

Dynamic programming (DP) is a way of solving decision problems by finding an optimal strategy. It thus falls into the collection of techniques used by operational research (OR) scientists and management scientists in their work. It provides a way of dealing with problems for which no other approach is as efficient. The types of problems with which DP is concerned are those described in outline above: problems which can be broken down into a sequence of single decisions made one after the other and problems where several decisions can be separated to give this same structure (here the word "after" is being used in its sense of both time and space).

Dynamic programming problems arise from many areas of commercial life. There are applications to the physical sciences and in sectors of industry where profit and financial objectives are not the overriding priority. Control rules for regulating service industries have been derived using DP; so have strategies for several sports and puzzles. Later in the book, we shall return to some of these areas of application.

The ideas at the root of dynamic programming were put forward by Richard Bellman in two books (Bellman (1957) and Bellman & Dreyfus (1962)) and several research papers. In the years since then there has been a constant stream of more books and papers which discuss the use of the ideas from both a practical and a theoretical point of view. The literature, especially in the mathematical development of the subject, is immense. Yet there are still unsolved problems, and as is the case with many areas of operational research, new areas of application appear regularly.

Dynamic programming differs from many of the techniques of OR in that there is no universal algorithm which can be applied to all problems. It is not like the more familiar forms of programming (which have little to do with computer programming) such as linear and integer programming. For these, there are sets of rules which can be applied to almost any problem and which will (sooner or later) find the optimal set of values of the parameters if one exists. Dynamic programming stands on both sides of the division between general-purpose and special-purpose OR models. Some attempts have been made to produce "black-box" solutions to DP problems, but these have not led very far. It is best seen as an approach to problem-solving, which requires thoughtful use, rather than as a technique which can be used by anyone who can read an instruction manual and feed in the numbers that are requested in the correct space and at the correct time.

Each dynamic programming problem must be formulated afresh, although guidelines have been established to help in this process for a number of problems. So inevitably, almost all DP problems need individual specification. At the same time, the rules which have been followed to create a particular formulation are general and can be recognized as being the "dynamic programming" method. In the same way as in most operational research studies, the problem being solved may be thought of as a system, described in a convenient way by a model. The model will include enough detail to be realistic without being unnecessarily complicated. There will be some constraints on the possible values of the decisions; there will be some way of relating the decisions which are taken to measurable and recognizable consequences.

1.4 The principles of dynamic programming

Any problem which requires the identification of the optimum of a function in N dimensions can be expressed in a standard form:

$$\text{minimise (or maximise)} \qquad f(x_1, \ldots, x_N)$$

subject to constraints on the values of $x_1, \ldots, x_N$. The list of functions and types of problems which can give rise to this format is a long one; linear programming, calculus, geometric programming, network optimisation, and so on .... In dynamic programming we shall alter this, from being one problem with N variables whose values we try and find simultaneously, to a succession of problems each associated with one of N **stages**. In each of these stages we will have problems to be solved with only 1 variable. Then we try to find the best value of a particular decision variable for that stage. Because it will not be possible to know the consequences of the other $N-1$ decisions, it will generally be essential to find the best value for decision variables for several different **states** (see below) at each stage. This usually means that we will have to solve more than the minimum of N problems; this is normally no hardship because it is generally easier to solve many problems with one variable than one problem with many variables. When solving each of these problems, we shall need to assume that the other variables have taken different possible values, or that different amounts of some resource are available when the decision is made at the stage being looked at.

This change to the form of a problem is usually referred to as being a **decomposition** of it. Often the decomposition will be a logical one, because there is a natural sequence of variables. Or the decomposition may be an artificial one selected by the person who is solving the problem. Whichever is the case the dynamic programming method is to solve the problems in the order that decomposition gives. Usually this will mean gradually increasing difficulty; in nearly every case, one problem will be expressed in a recursive form involving others whose solutions have been found.

It is often convenient to consider all dynamic programming problems as being sequential in time for descriptive ease. Then each variable corresponds to a decision made at a specific epoch or moment in time. It is a natural consequence of this that, after a decision has been made, the system being modeled will change to a new **state** in advance of the next stage in time. The change between states at successive stages depends on the decision, the stage and the state.

Therefore, over time, the system will progress through a succession: stage in time in a particular state, first decision, second stage in time in a new state, second decision, and so on until the final (the Nth) decision and the final state. The sequence of N decisions which helps determine this progress through time is a **policy** for the model. This policy and the associated sequence of N states of the model will yield a value for the function (of N variables) being optimised. Thus it would be possible to find the optimal policy by trial and error over all the possible policies.

A simple example of a sequential decision problem was given earlier in this chapter; how should production of an item be managed, where the costs of production and storage vary, and the demand is random? At the start of the problem, the "system" had 4 items; a decision had to be made about the number (say X_1) to be produced during the first week, and the random demand (D_1) during that week would mean that the state of the system at the beginning of the second week would be $4 + X_1 - D_1$. Then the second decision will be made; the random demand of the second week will mean that the decision-maker will start the third week in a new state; and similarly in the fourth week. The succession is evident. It is: stage in time "week 1", in state "4 items", decision "X_1" ... leading to "week 2", state "$4 + X_1 - D_1$" items, ... One approach to solving the problem would be to calculate what the total costs of managing this small company for four weeks would be, with all the possible decisions that might be made at the start of each week, and all the possible random demands that could occur.

However, instead of such brute force methods of total enumeration, dynamic programming relies on a principle, the so-called **principle of optimality**, which facilitates the identification of optimal policies.

The principle of optimality (also known as Bellman's principle) can be expressed in many different ways. If we want to know the best decision which can be made from a given state and stage of the problem, we must consider each decision and each state which that decision would lead to (at the next stage). But we do not need to go any further than the next stage; after that we will follow an optimal policy. So the optimal policy can be found by comparing the outcomes from each decision and the optimal policy from the resulting states. Once this has been done, optimal policies for earlier states can be found. Some of the ways that this principle can be formally enunciated are:

> An optimal policy has the property that whatever the state the system is in at a particular stage, and whatever the decision taken in that state, then the resulting decisions are optimal for the subsequent state.

or

> An optimal policy is made up of optimal subpolicies.

or

> An optimal policy from any state is independent of how that state was achieved and comprises optimal subpolicies from then on.

In many ways the principle of optimality follows from common sense and it can be proved intuitively by contradiction. For if the system is in state A at some stage and the optimal policy takes it to state B at the next stage, then the optimal subpolicy from B must coincide with the corresponding part of the optimal policy from A. If not, then one or the other could be improved. In the example above, the optimal policy for four weeks of managing the production of the items might include the possibility of starting the third week with 3 items. The principle of optimality says that the best thing to do if this happens is to follow the best policy for managing the production for the last two weeks, with 3 items as the opening inventory in the first of them. In other words, one can forget that the 3 items might be the result of managing the production for four weeks.

One of the variant forms of the principle of optimality implied that an optimal policy is independent of the past and only looks to the future. This is an essential part of dynamic programming, as it allows policies to be calculated recursively without the need for modification. It does mean that the identification of the state must be sufficient to fully describe the system so that this independence may be observed, and so that the problem can be effectively decomposed into a series of one-dimensional problems each of which depends on the solution to later problems (optimal subpolicies for subsequent states) but not on the solution to earlier ones. So, in the example of the last paragraph, the optimal policy for the last two weeks starting with an inventory of 3 items does not depend on the orders in the weeks before that.

In these assumptions about optimality, there is no specific statement about the mathematical form that is used for the calculation of the objective function. This is deliberate. No such statement is necessary. In later chapters we shall explore several distinct expressions, each of which is a special purpose model for a particular problem. The common feature in each formulation or expression is a **recurrence relationship**, which links the optimal objective function value in one state and stage with that corresponding to the next stage and one or more possible states of it. In this relationship we shall use the costs of passing between states (or the cost of making a decision). (Cost is a convenient term for the units of the objective function; it is possible to think of objective functions whose value may be expressed in any units ...it is simply a handy way of thinking about some kind of unit.) The form of the recurrence relationship will depend on many aspects of the model, among which is whether or not the model has stochastic (random) aspects or not: there is no uniformity of functional form.

One consequence of the form of dynamic programmes is that, for some or all the stages, there may be a large number of possible states. To find the best set of decisions means that we shall need to calculate the optimal action in all the possible states at each stage. This may lead to a very large number of calculations, and it might be thought that a method which leads to such a burden of work could not be particularly useful in practice. In most situations where there are numerous states, the calculations involved are comparatively simple ones, so that they can be completed reasonably quickly. They will also (in general) be easier than the corresponding problems without the dynamic programming formulation. As we shall see later, there are situations where there are computational difficulties associated with having models where the number of states that might occur is very large.

2

Simple Recursive Problems

2.1 The stagecoach problem

"Once upon a time ..." is the traditional way that all stories begin. In dynamic programming, the simplest problem can be easily illustrated by a fictitious account which also begins in this way. The story goes that a businessman wished to travel by stagecoach in the wild west of North America in the middle of the 19th century. Stagecoaches ran from state capital to state capital; these were the only places where it was possible to change stagecoaches. Each part of the journey had risks from the thieves and the other perils of travel at this period of history. Prudent travellers would buy insurance policies to protect their lives and property; the insurers calculated their premiums per stagecoach journey on the basis of the known or estimated risks compared with other routes. Needless to say, a prudent traveller would try to find the route which boasted the lowest total premium. This would save money and – assuming that the vendors of insurance were honest – be the safest route to follow. The businessman in the story found that his choice of stagecoach journeys could be represented by the diagram (Figure 2.1) and that the premiums were those shown in the table (Table 2.1). The story asks for the best route from the start to the finish of a proposed journey. (One could imagine a similar journey in the twentieth century, with the insurance costs replaced by the fares for different forms of public transport.)

To find the cost of an insurance policy for the journey we simply add together the separate costs of each premium ... there are no discounts. In a small example such as this, it is simple to calculate the costs of all possible policies (there

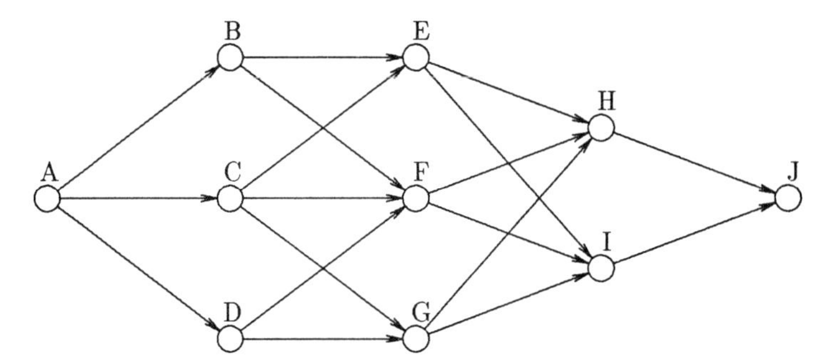

Figure 2.1: Possible routes for the businessman. Letters A – J represent the state capitals; the proposed journey starts at A and ends at J

are 14 in all) and identify the cheapest. But it is helpful to see how such a problem (which is more or less manageable by hand) can be solved by dynamic programming. The problem could also be treated as a shortest path problem and solved by one of the family of methods which exist for network optimisation.

Table 2.1: Insurance premiums

From To	B	C	D	E	F	G	H	I	J
A	22	19	15						
B				16	15				
C				22	16	22			
D					19	22			
E							16	18	
F							18	11	
G							15	13	
H									23
I									29

The first part of the dynamic programming process is to state the objective. In this case we want to identify a route through the network which will minimise the total cost of the insurance policy. So the optimal policy (in terms of DP) is the cheapest policy (in terms of insurance).

The problem naturally appears as a series of decisions taken over time (as the journeys between state capitals take some days to complete) or over space (as these capitals are separated by some distance). In each state (capital) that he visits, the businessman has a choice of stagecoach journeys to make, and he must decide on one of them. Whichever route he follows from the start to the finish of the trip, there will be four such decisions, some of which may be simple ones (where there is no choice). The journey will be made up of four stages (stagecoaches). So the problem is that of finding the best policy for a decision problem involving four stages and a finite (one, two or three) number of states possible at each stage. The problem has been specially devised to have the close links with the key expressions of all dynamic programming problems. (The stagecoach problem with these direct parallels to the nomenclature of dynamic programming, was developed by Professor H. M. Wagner and has appeared in print in several textbooks. Among these is Wagner's text(1975), where the fictitious businessman is referred to as Mark Off.)

The recurrence relation which defines the cost of a policy is a simple one. If $f_n^*(i_n)$ is the cost of an optimal policy when there are n stages remaining, and the decision is made in state i_n, then

$$f_n^*(i_n) = \min_{i_{n-1}}\{p(i_n, i_{n-1}) + f_{n-1}^*(i_{n-1})\} \tag{2.01}$$

where $p(a,b)$ is the premium associated with the single stagecoach journey from a to b. We define $f_0^*(J) = 0$ to indicate the completion of the journey. The notation $\min_j$ will be used to indicate the minimum of a function taken over all the possible values of a variable j; this range of permissible values will depend both on the context of the problem and on constraints.

This recurrence relationship may be expressed informally: the least cost of a policy from state i_n, with n stages of the journey left, is found by listing the costs of the journeys to all destinations which are possible using one stage from i_n and then adding the cost of an optimal policy from each destination; then selecting the least expensive on the list.

And so we can solve the problem with the data that has been given.

When there is *one* stage left, there are two possible states for the businessman, H and I. In each one he has no choice of destinations; he must go to J directly. Thus:

$$f_1^*(H) = 23 + f_0^*(J) = 23 + 0 = 23 \tag{2.02}$$

and similarly

$$f_1^*(I) = 29 + f_0^*(J) = 29 + 0 = 29 \tag{2.03}$$

When there are *two* stages left, the businessman can be in any of the three states E, F and G. In each one he has a choice of two destinations, H and I. If he goes from E to H, the cost of this stage is 16 and then he follows the optimal policy from H to J; if he goes from E to I, then he will pay 18 for the single stage, followed by the cost of the optimal policy from I to J. So we can calculate:

$$\begin{aligned} f_2^*(E) &= \min(16 + f_1^*(H), 18 + f_1^*(I)) \\ &= \min(16 + 23, 18 + 29) \\ &= 39 \text{ (corresponding to deciding to go to H)} \end{aligned} \tag{2.04}$$

Similarly

$$\begin{aligned} f_2^*(F) &= \min(18 + f_1^*(H), 11 + f_1^*(I)) \\ &= \min(18 + 23, 11 + 29) \\ &= 40 \text{ (corresponding to deciding to go to I)} \end{aligned} \tag{2.05}$$

and:

$$\begin{aligned} f_2^*(G) &= \min(15 + f_1^*(H), 13 + f_1^*(I)) \\ &= \min(15 + 23, 13 + 29) \\ &= 38 \text{ (corresponding to deciding to go to H)} \end{aligned} \tag{2.06}$$

One stage earlier, there are *three* stages remaining, and the businessman has three states in which he could find himself, B, C and D. Now the choice of routes is restricted according to the current state; it is neither possible to go from B to G nor from D to E. Using the recurrence relation again, we find that:

$$\begin{aligned} f_3^*(B) &= \min(16 + f_2^*(E), 15 + f_2^*(F)) \\ &= \min(16 + 39, 15 + 40) \\ &= 55 \text{ (it does not matter which decision is made)} \end{aligned} \tag{2.07}$$

$$\begin{aligned} f_3^*(C) &= \min(22 + f_2^*(E), 16 + f_2^*(F), 22 + f_2^*(G)) \\ &= \min(22 + 39, 16 + 40, 22 + 38) \\ &= 56 \text{ (corresponding to deciding to go to F)} \end{aligned} \tag{2.08}$$

$$\begin{aligned} f_3^*(D) &= \min(19 + f_2^*(F), 22 + f_2^*(G)) \\ &= \min(19 + 40, 22 + 38) \\ &= 59 \text{ (corresponding to deciding to go to F)} \end{aligned} \tag{2.09}$$

One stage earlier, with *four* stages remaining, the only state corresponds to the start of the journey and so the optimum policy from this state will be the one which is desired in the problem. The businessman has a choice of three routes. These lead to a recurrence relationship:

$$\begin{aligned} f_4^*(A) &= \min(22 + f_3^*(B), 19 + f_3^*(C), 15 + f_3^*(D)) \\ &= \min(22 + 55, 19 + 56, 15 + 59) \\ &= 74 \text{ (corresponding to deciding to go to D)} \end{aligned} \tag{2.10}$$

So we have found the optimum policy for the whole journey. The first decision will be to go to D; from D the optimal subpolicy (three stages) is to go to F (2.09); the optimal subpolicy from F is to go to I (2.05), and from there the optimal subpolicy (which was obvious) was to go straight to J. Hence the route is A–D–F–I–J with a total cost of 74 units.

By the principle of optimality, all parts of this policy are optimal for the subproblems that they represent. In constructing the solution we used the optimality of the path from I to J, that from F to J and that from D to J. In addition to these optimal subpolicies, we have also found the optimal subpolicies for paths from A to I, from A to F and from D to I. The calculations which were completed but not actually used in the optimal policy (that from A to J) yield several other optimal subpolicies: B to H, B to I, B to J, C to I, C to J, E to J and G to J.

2.2 Sensitivity analysis of the shortest route

Other questions naturally arise in connection with a problem such as this. In operational research one often wishes to know about the sensitivity of a solution to the data which have been used in it. Interest of this kind is often posed as a "What if ... ?" question. "What if the premiums on one stage are raised by three dollars?" ... "What if it is impossible to visit a given state?" ... "What if the businessman insists on visiting a given state?" and so on. We might like to know whether such changes affect the optimal policy and if so, by how much the policy cost will be changed.

Examination of the form of the solution is a great help toward the answering of such questions. If a particular stage has its premium changed then the cost of policies will change if they use that stage. The cost of policies from states preceding that stage may change as a result. So may the optimal subpolicies that straddle the selected stage. We may make some simple observations without recourse to mathematics. A reduction in premium for any stage in the optimal policy will only affect the cost of the policy, not the policy itself. An increase in the premium on any stage which is not on the optimal policy will affect neither the cost nor the policy. But an increase in the premium for a stage which is part of the optimal policy will affect the cost of the optimal policy, and will possibly change the optimal policy itself when the increase makes some other policy better. Conversely, a decrease in a premium on any stage which is not on the optimal policy will only affect the optimal policy when the changed cost is sufficiently large.

The need to avoid a particular state is a particular case of changed premiums. If one cannot (or does not wish to) visit a given state, then the effect will be the same as changing the premiums of all stages which start or finish at that state by some very large penalty. Thus the same reasoning as was described above will be valid; the optimal policy will only be changed if it included the forbidden state.

For an optimal policy which includes a specified state, S say, there are two obvious approaches: either one can reduce the premiums on all stage routes which include that state by some large amount and solve the problem using the methodology we have already used (finding the cost with the optimal route rather than the optimal values of $f_4^*(A)$; alternatively, one can find the optimal subpolicy from S to the destination (J) and the optimal subpolicy from the start (A) to S.

2.3 The general shortest route problem

The stagecoach problem is artificial, but is a useful way of indicating the dynamic programming approach to problems. In most shortest route problems there will not be the same neat structure of states and stages; the number of arcs in the best route for a traveller may not be known in advance (we knew in the example that the businessman would need to take exactly four stages to complete his journey). In such cases the simplified approach we have used so far will not work. Instead, we can use a more general method. Suppose that the network has N nodes and that they are numbered so that we are looking for the shortest route from node 1 to node N. It should be immediately apparent that the best route will pass through at most $N-2$ intermediate nodes and will use at most $N-1$ arcs. Otherwise one or more nodes will be visited more than once, and this would be an absurd situation. This simple observation means that we can define a suitable objective function to be minimised in the recurrence relationship of dynamic programming. We will need to consider a variety of shortest routes in the network, which will vary according to where they start and the maximum number of arcs that will be used to get to node N. So, the state of the system can be defined, as before, by the node one requires the shortest route to node N from (that is, the state one is in, when taking the decision). The stages can be numbered according to the largest possible number of arcs in the route from there to node N. Thus we identify $f_n^*(i)$ as the length of the shortest route from node i to node N using n arcs or less. The answer to our problem will be (in this notation) $f_{N-1}^*(1)$. The recurrence relationship to use will be the same as with the stagecoach problem, that is:

$$f_n^*(i) = \min_j \{p(i,j) + f_{n-1}^*(j)\} \tag{2.11}$$

where $p(a,b)$ is the distance from node a to node b. There will be an optimal destination for each state at each stage; the value of j which is selected by equation (2.11), $j = j_n^*(i)$ so that:

$$f_n^*(i) = p(i, j_n^*(i)) + f_{n-1}^*(j_n^*(i)) \tag{2.12}$$

The recalculation will stop when $f^*_{N-1}(1)$ has been calculated, or when $f^*_n(i) = f^*_{n-1}(i) \forall i$. This second possibility indicates that there is no chance of improving $f()$. To start off the recurrence relationship, we will need to specify the values of $f^*_0(i)$, and this can be done easily with

$$f^*_0(N) = 0$$
$$f^*_0(i) = \infty \quad (i \neq N) \tag{2.13}$$

There is one important difference; in the stagecoach problem we assumed that the decision taken in a particular state would lead to a new, different state. Here, in the general problem, this need not be the case. We may choose to stay in the same state, and we shall take $p(i,i) = 0$ for all i. Such an assumption is necessary because we do not want to force the optimal route from a given state to use a fixed number of arcs. To prevent ambiguity, it is usually desirable that the only state in which it is possible to stay is the destination, node N.

There is also an implied assumption that the cost $p(i,j)$ is non-negative for all the arcs. When $p(i,j)$ represents a distance, a consumption of fuel or a time, this is reasonable. However, $p(i,j)$ may represent a cost, and it is possible for this to be negative on a particular arc. This would be the case if there was some payment for using this arc. In such circumstances, the same formulation as given earlier can be used, but the existence of a solution depends on whether or not there are any circuits in the network which have a total cost which is negative. Such circuits could be repeated infinitely, each time reducing the cost still further. To guard against this happening is difficult because such circuits may not be apparent in the data which is presented; in solving a problem by hand they may appear and be recognizable. Using a computer we can find circuits by calculating $f^*_N(i)$, $f^*_{N+1}(i) \ldots f^*_{2N-2}(i)$ and checking that these all have the same values for all i.

2.4 Example of a general shortest route

Consider the network shown in Figure 2.2 where the costs $p(i,j)$ are shown in Table 2.2; what is the shortest route from node 1 to node 7?

Table 2.2: Distances for the network of figure 2.2

	2	3	4	5	6	7
1	12	14	24	25		
2		7		3		14
3			-5		8	
4				-3	4	5
5						3
6						3

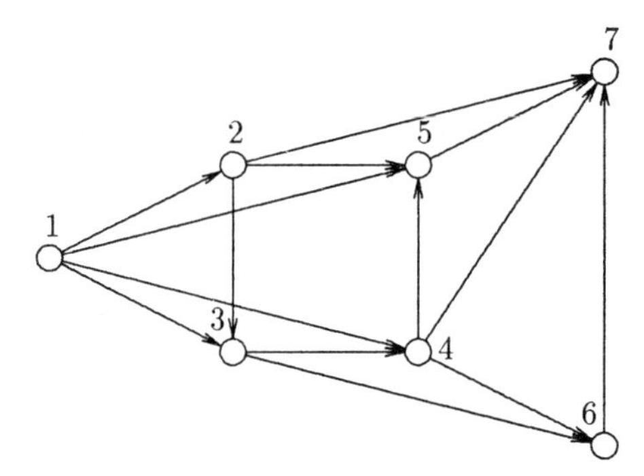

Figure 2.2: Network for a shortest path

Table 2.3: Calculations required to find the shortest paths to node 7

$i \rightarrow$	1	2	3	4	5	6	7
$f_0^*(i)$	∞	∞	∞	∞	∞	∞	0
$f_1^*(i)$	∞	14	∞	5	3	3	0
$j_1^*(i)$		7		7	7	7	7
$f_2^*(i)$	26	6	0	0	3	3	0
$j_2^*(i)$	2	5	4	5	7	7	7
$f_3^*(i)$	14	6	-5	0	3	3	0
$j_3^*(i)$	3	5	4	5	7	7	7
$f_4^*(i)$	9	2	-5	0	3	3	0
$j_4^*(i)$	3	3	4	5	7	7	7
$f_5^*(i)$	9	2	-5	0	3	3	0
$j_5^*(i)$	3	3	4	5	7	7	7

Successive values of f_n^* and j_n^* can be found by using the recurrence relationship and these are tabulated in Table 2.3. It will be seen that $f_4^* \equiv f_5^*$. Hence, the solution has been found, although the recurrence relation would automatically have terminated when f_6^* had been evaluated.

2.5 Variants of the shortest route problem

Puzzles for children and competitions for adults often involve problems of routes. A maze is a simple example, where the purpose is to find a route through a network which has been drawn in a disguised way so as to confuse the puzzler. Rearranging the maze diagram is usually enough to solve the problem, but it removes the element of fun and enjoyment from it.

Another form of route finding is that shown in Figure 2.3. In this figure, we have five columns and five rows of positive numbers, one in each of 25 cells. A "route" through the figure consists of moving from a cell in the leftmost column to a cell in the rightmost column, using one cell from each column and always moving from a cell to one which is either adjacent or touching at a corner. The objective is to find a route through the cells so as to optimise some function of the scores in the cells which are visited. Any cell in the first column can be used as a starting point, and the final cell is also arbitrary.

	col 5	col 4	col 3	col 2	col 1
Row A	28	45	18	45	32
Row B	5	23	31	20	9
Row C	38	16	48	23	28
Row D	34	3	27	34	19
Row E	22	38	11	47	31

Figure 2.3: Finding a route through cells

Suppose the data in Figure 2.3 are applied to a problem of finding a route from column 5 to column 1 so that the total of the scores in the five cells which are visited is as large as possible. We denote the scores as $d(r, c)$ for the value of the cell in row r and column c (r = A, B, C, D, E; c = 1, 2, 3, 4, 5). By analogy with the stagecoach problem, the problem can be described according to the row (state) and column (stage) which has been reached, and the optimal policy from there will be the one which maximises the score.

So a recurrence relation may be deduced:

$$f_n^*(r) = \max_{row\ y\ adjacent\ to\ row\ r} \{d(r,n) + f_{n-1}^*(y)\} \tag{2.14}$$

with

$$f_0^*(r) = 0 \qquad \forall r \tag{2.15}$$

Then

$$f_1^*(A) = 32 \quad f_1^*(B) = 9 \quad f_1^*(C) = 28 \quad f_1^*(D) = 19 \quad f_1^*(E) = 31$$

$$\begin{aligned}
f_2^*(A) &= \max(45 + 32, 45 + 9) = 77 \\
f_2^*(B) &= \max(20 + 32, 20 + 9, 20 + 28) = 52 \\
f_2^*(C) &= \max(23 + 9, 23 + 28, 23 + 19) = 51 \\
f_2^*(D) &= \max(34 + 28, 34 + 19, 34 + 31) = 65 \\
f_2^*(E) &= \max(47 + 19, 47 + 31) = 78
\end{aligned}$$

$$f_3^*(A) = \max(18 + 77, 18 + 52) = 95$$
$$f_3^*(B) = \max(31 + 77, 31 + 52, 31 + 51) = 108$$
$$f_3^*(C) = \max(48 + 52, 48 + 51, 48 + 65) = 113$$
$$f_3^*(D) = \max(27 + 51, 27 + 65, 27 + 78) = 105$$
$$f_3^*(E) = \max(11 + 65, 11 + 78) = 89$$

And:

$$f_4^*(A) = 153 \quad f_4^*(B) = 136 \quad f_4^*(C) = 129 \quad f_4^*(D) = 139 \quad f_4^*(E) = 143$$

Finally:

$$f_5^*(A) = 181 \quad f_5^*(B) = 158 \quad f_5^*(C) = 177 \quad f_5^*(D) = 177 \quad f_5^*(E) = 166$$

This indicates that the best cell to start from is in row A, and then we may deduce that the route passes through (A,5), (A,4), (B,3), (A,2), (A,1). This is in spite of the unattractiveness of (A,5) as the initial starting point.

The same grid and numbers could be used in other puzzles, with different criteria for optimality. One could search for the path whose score had the smallest total value, in which case the recurrence relationship would be

$$f_n^*(r) = \min_{row\ y\ adjacent\ to\ row\ r} \{d(r,n) + f_{n-1}^*(y)\} \tag{2.16}$$

with

$$f_0^*(r) = 0 \quad \forall r \tag{2.17}$$

or the path whose score to be maximised was found by taking the product of the individual values, yielding the recurrence relationship and starting conditions:

$$f_n^*(r) = \max_{row\ y\ adjacent\ to\ row\ r} \{d(r,n) \times f_{n-1}^*(y)\} \tag{2.18}$$

with

$$f_0^*(r) = 1 \quad \forall r \tag{2.19}$$

Furthermore, such "puzzles" need not be restricted to rectangular grids similar to Figure 2.3. Provided that the states, stages and recurrence relationships can be defined, then the approach indicated in this section can be used. Frequently, with grids other than a rectangular one, the problem is one of deciding which states can be reached. There have been some practical applications of this type of formulation to decisions about the price to charge for goods. When the demand is sensitive to the price, a rapid variation in the retail price may be bad for business. Market research may indicate the expected profit that could be achieved from varying the price at different times and this will provide a table of revenue similar to that of Figure 2.3. In such a formulation the columns represent the times that the price could be changed, and the rows the possible levels of price.

2.6 Approximate methods in shortest routes

The solution method for the general shortest route problem is a little tedious to use and very often a good policy may be apparent without calculation ... simply using common sense. The recurrence relationship may then be used to test whether or not the apparent solution is optimal, using the following: the solution to the recurrence relationship was a function $f^*_{N-1}(i)$, the route using at most $N-1$ links from i to N. Clearly $f^*_{N-1}(i)$ is equal to $p(i,j) + f^*_{N-2}(j)$ for some j. For this node j we have $f^*_{N-1}(j) = f^*_{N-2}(j)$. (Otherwise the optimal route from j to N would require $N-1$ links and this route would have to pass through node i, giving an optimal route from i which went to j and then later returned to i which would be a contradiction. So $f^*_{N-1}(i) = p(i,j) + f^*_{N-1}(j)$ for some j. We simplify this by writing $f^*(i)$ in place of $f^*_{N-1}(i)$ and work with this function whose value is the shortest distance from i to N. Then:

$$f^*(i) = p(i, j^*_i) + f^*(j^*_i) \tag{2.20}$$

$$\text{solves } f^*(i) = \min_j\big(p(i,j) + f^*(j)\big) \tag{2.21}$$

A policy may be guessed at, identifying guessed destinations j_i for each node i, and with these, the corresponding values of $f(i)$ can be found. These values may not be optimal (and so they have not been given a $*$ superscript). It is essential that these values j_i are self-consistent. So we require that for each i there is a route to node n. Then the recurrence relationship can be used to improve each f in turn.

The application of this rule will consist of the solution of N simultaneous equations in N unknowns $f(i)$ $(i = 1, 2, \ldots, N)$ (which will be very simple to do as each one involves two terms only). This will be followed by use of the rule (2.21) above to recalculate $f(i)$ and possibly reduce its value $i = 1, 2, \ldots, N, 1, 2, \ldots, N, 1, \ldots$. The iterations stop when N applications of the rule do not change the value of the optimal policy. Doing this may well speed up the calculations. The gain depends on the accuracy of the initial guesses, as well as the sequence of the nodes. If the first guess is j_i =N for all i, then the iterations are effectively those that follow from the original dynamic programming formulation.

2.7 Example of approximation

In the example described earlier, the best policy corresponded to:

i	1	2	3	4	5	6	7
j^*_i	3	3	4	5	7	7	7

If we had made a guess which was close to this, such as:

i	1	2	3	4	5	6	7
j_i	2	3	4	7	7	7	7

then the approximate method could be used to find that, with these values of j_i, the function would be the solution to:

$$f(1) = 12 + f(2) \quad f(2) = 7 + f(3) \quad f(3) = -5 + f(4) \quad f(4) = 5 + f(7)$$
$$f(5) = 3 + f(7) \quad f(6) = 3 + f(7) \quad f(7) = 0 + f(7)$$

and since $f(7) = 0$, these yield:

$$f(1) = 19 \quad f(2) = 7 \quad f(3) = 0 \quad f(4) = 5 \quad f(5) = 3 \quad f(6) = 3 \quad f(7) = 0$$

Using the iterative rule:

$f(1) = \min\{12 + 7, 14 + 0\}$
$= 14$ (corresponding to a changed policy with $j_1 = 3$)
$f(2) = \min\{7 + 0, 3 + 3, 14 + 0\} = 7$ (no change)
$f(3) = \min\{-5 + 5, 3 + 3\} = 0$ (no change)
$f(4) = \min\{-3 + 3, 4 + 3, 5 + 0\}$
$= 0$ (corresponding to a changed policy with $j_4 = 5$)
$f(5) = 3$ (no change)
$f(6) = 3$ (no change)
$f(7) = 0$ (no change)

$f(1) = \min\{12 + 7, 14 + 0\} = 14$ (no change)
$f(2) = \min\{7 + 0, 3 + 3, 14 + 0\} = 7$ (no change)
$f(3) = \min\{-5 + 0, 3 + 3\}$
$= -5$ (a changed value but no change in policy)
$f(4) = \min\{-3 + 3, 4 + 3, 5 + 0\} = 0$ (no change)
$f(5) = 3 \quad f(6) = 3 \quad f(7) = 0$

$f(1) = \min\{12 + 7, 14 - 5\}$
$= 9$ (a changed value but no change in policy)
$f(2) = \min\{7 - 5, 3 + 3, 14 + 0\}$
$= 2$ (a changed value but no change in policy)
$f(3) = \min\{-5 + 0, 3 + 3\} = -5$ (no change)
$f(4) = \min\{-3 + 3, 4 + 3, 5 + 0\} = 0$ (no change)
$f(5) = 3 \quad f(6) = 3 \quad f(7) = 0$
$f(1) = \min\{12 + 2, 14 - 5\} = 9$ (no change)
$f(2) = \min\{7 - 5, 3 + 3, 14 + 0\} = 2$ (no change)

As the last $N = 7$ calculations have not changed the policy or its value, the optimal solution has been found. The amount of work required to find this has been a little less than the direct method entailed. If the initial policy for $i = 4$ had been guessed more accurately, then we would have saved work; the error we made in the guess had to be put right and the correction worked through the system. (In some cases, effort could be saved by recalculating the f values after each set of N evaluations to find the j_i values.)

2.8 Summary of chapter 2

In this chapter we have examined the use of dynamic programming for finding routes through networks. We have progressed from the simple (and artificial) problem of stagecoaches to more general routing problems, and to the use of approximate methods. In some ways many examples of dynamic programming draw their inspiration from the idea of finding a path through a series of states, with one decision being taken at a time. The stages are the successive times when these decisions are taken.

2.9 Exercises

(1) Solve the stagecoach problem of Table 2.4.

Table 2.4: Stagecoach problem (exercise 1)

From To	B	C	D	E	F	G	H	I	J	K
A	14	11	12							
B				17	25					
C				20	24	20				
D					25	14				
E							21	24		
F							15	15	21	
G								12	18	
H										10
I										24
J										13

	col 6	col 5	col 4	col 3	col 2	col 1
Row A	25	37	12	35	21	33
Row B	23	36	13	34	13	34
Row C	22	35	14	35	11	36
Row D	20	36	11	32	13	37
Row E	24	38	12	32	11	36

Figure 2.4: Finding a route through cells (exercise 2)

(2) Using dynamic programming, solve the problem in Figure 2.4 with the objective of minimising the product of the scores. (All other details are as in the example illustrated in Figure 2.3)

(3) Solve the problem in Figure 2.2 with a new arc from node 6 to node 7 whose cost/distance is -10.

(4) Solve the stagecoach problem of Table 2.5.

Table 2.5: Stagecoach problem (exercise 4)

From To	B	C	D	E	F	G	H	I	J	K
A	-16	-11	-12							
B				21	-39					
C				-38	27	22				
D					20	-39				
E							47	43		
F							46	48	45	
G								46	45	
H										16
I										12
J										13

(5) In problems such as those shown in Figure 2.3, suppose it is desired to find a path for which the difference between the largest score and the smallest score is as small as possible. Consider whether this can be formulated as a dynamic programme. If it can, give a recurrence relation and solve the problem using the numbers in Figure 2.3. If it cannot be solved as a dynamic programme, explain why it is not possible.

(6) One of your young relatives, knowing that you are "clever with sums" has asked for your help with a puzzle from the back of her breakfast cereal packet. Galahad, the hero of the story, has to take his herd of white unicorns to one of the markets shown at the right hand side of Figure 2.5; when he reaches one of these, the king will reward him with the number of gold coins shown beside the names of the towns. However, feeding and guarding unicorns costs money, and Galahad knows the cost of taking the herd through each of the possible intervening towns. This cost, also in gold coins, is shown beside each town. The arrowed lines show the routes that he can take between the towns. What route should Galahad follow?

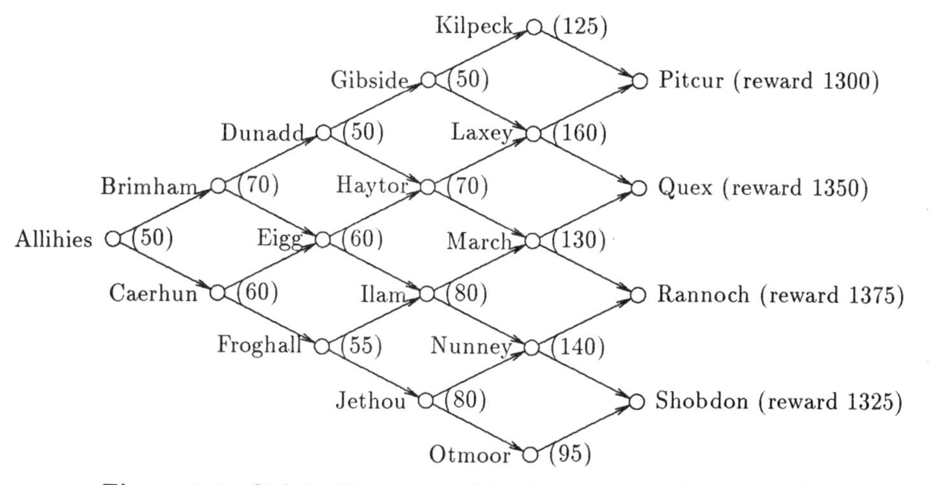

Figure 2.5: Galahad's route with his unicorns (exercise 6)

(7) Suppose that you are trying to find a path in a problem such as Figure 2.3, but your objective is that of minimising the variance of the scores, that is, you want to minimise

$$\frac{\left[\sum_{cells\ visited} X_{r,c}^2 - \left(\left(\sum_{cells\ visited} X_{r,c}\right)^2/n\right)\right]}{(n-1)}$$

(where n is the number of cells visited.)
Can this problem be formulated as a dynamic programme?

(8) The sequence of integers $X_1, X_2, \ldots, X_N$ contains both positive and negative values. Find a dynamic programming formulation which will solve the problem of finding integers I, J with $1 \leq I \leq J \leq N$ such that the sum

$$\sum_{k=I}^{k=J} X_k$$

is minimised. Demonstrate your algorithm on the sequence:

$$-1, 5, 0, 1, 8, -10, 14, -2, -6, 5$$

(9) A factory makes steel strip which is 2 cm thick. This is produced by processing thicker strip through three rolling machines from an input of 10 cm thick material. The machines are identical; each one takes input from the previous one; the speed of operation $s_{i,j}$ depends on the thickness of the input (i cm) and the thickness of the output (j cm) for that machine. These speeds are given in Table 2.6; the speed of the production process is determined by the speed of the slowest machine. The management wants to produce the steel strip as fast as possible. Use dynamic programming to find the settings on each machine and the optimal speed of production. (In the table "n.a." means that a particular setting is not available.)

What difference would there be if the input strip were to be only 9 cm thick?

Table 2.6: Speed of rolling presses (exercise 9)

	Output								
Input	2 cm	3 cm	4 cm	5 cm	6 cm	7 cm	8 cm	9 cm	10 cm
2 cm	1.00	n.a.	n.a.	n.a.	n.a.	n.a.	n.a.	n.a.	n.a.
3 cm	0.80	1.00	n.a.	n.a.	n.a.	n.a.	n.a.	n.a.	n.a.
4 cm	0.70	0.85	1.00	n.a.	n.a.	n.a.	n.a.	n.a.	n.a.
5 cm	0.63	0.76	0.88	1.00	n.a.	n.a.	n.a.	n.a.	n.a.
6 cm	0.57	0.70	0.80	0.91	1.00	n.a.	n.a.	n.a.	n.a.
7 cm	n.a.	0.62	0.74	0.83	0.92	1.00	n.a.	n.a.	n.a.
8 cm	n.a.	n.a.	0.69	0.76	0.84	0.92	1.00	n.a.	n.a.
9 cm	n.a.	n.a.	n.a.	0.68	0.79	0.84	0.91	1.00	n.a.
10 cm	n.a.	n.a.	n.a.	n.a.	0.70	0.75	0.83	0.93	1.00

Hint: Think of the process as illustrated in the three diagrams in Figure 2.6: in the first, strip of an unknown thickness X_1 cm is input into the last machine, and must emerge at thickness 2 cm; in the second, strip of unknown thickness X_2 cm enters the second to last machine and emerges at thickness X_1 cm – a decision variable. The speed of production of 2 cm thick steel is the smaller of two speeds; that on the second to last machine, where it is reduced from X_2 cm to X_1 cm, or the speed of the remaining machines which reduce it from X_1 cm

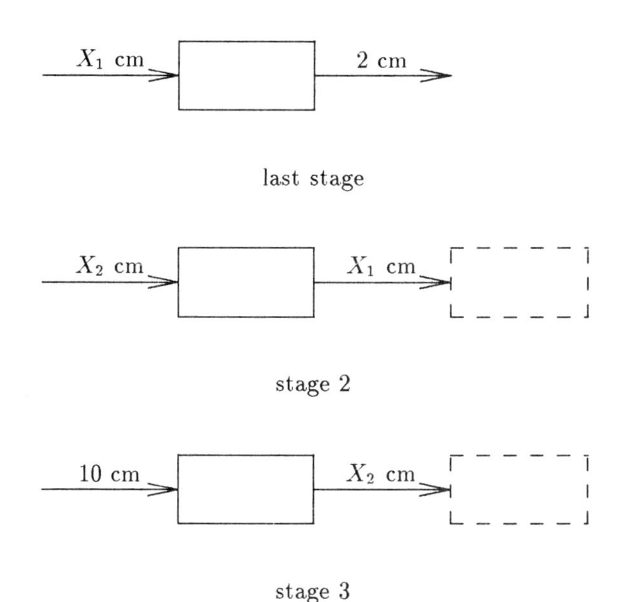

Figure 2.6: The sequential nature of the rolling presses

to 2 cm. (We don't need to know how many machines are involved in the process of rolling it from X_1 cm to 2 cm; we simply need the speed.)

In the third diagram, strip of thickness X_3 cm (and the question is interested in $X_3 = 9$, $X_3 = 10$) is reduced to thickness X_2 cm and then proceeds to the remaining machines. The speed of production is the smaller of the speed on the third to last machine and the speed of the "system" that follows (and again, we don't need to know how many machines are involved in this "system").

Table 2.7: Pages in a stamp collection (exercise 10)

Letter	Pages	Letter	Pages	Letter	Pages
A	47	B	38	C	55
D	4	F	6	G	54
H	2	I	42	J	8
K	13	L	2	M	77
N	49	O	1	P	25
Q	2	R	2	S	73
T	51	U	3	V	4
W	1	Z	8	total	567

(10) My stamp collection is housed in five stamp albums, each of which is designed to hold up to 150 pages. For convenience, I want each album to hold the pages that correspond to all the countries whose names begin with consecutive letters of the alphabet. Thus album number 1 should contain letter "A", or letters "A" and "B", or letters "A", "B" and "C", and so on; album number 2 should follow on in sequence.
Ideally, the maximum number of pages in any of the albums should be as small as possible. At present, my collection has contents given in Table 2.7. (I do not collect stamps from countries whose names begin with "E", "X" and "Y".)
Formulate the problem of finding the best arrangement of pages in my stamp albums as a dynamic programme.
Naturally, from time to time, my stamp collection expands, and new pages are added. How would you extend the formulation to indicate when I will need to buy a new album?

3

Knapsack Problems

3.1 Introduction

When you travel on holiday or on business, there is usually some limit to the amount of luggage which you can take with you. If you are hiking, then your backpack will only hold a limited weight and volume of items. If you are flying by scheduled airline, there are limits to the weight (or possibly, the volume) of luggage that you can carry without incurring any surcharge. These depend on the route you are taking and on whether you are travelling first class or not. (If you choose to travel by an unscheduled flight, in a private aeroplane, there will be similar restrictions. These are not so clearly established as the rules for scheduled flights.) Should you be going by road or rail, then you will be limited by the amount of material that can be easily carried by whatever vehicle you have decided to use. As a result of these restrictions you must make a decision about which items to take ... and which to leave behind! Ideally, each part of your luggage should be worth having with you (or at least, you should be reassured to know that you have it with you) at some stage in your journey, and should be worth more than the items that you could have packed but left behind because of the size of your trunk or suitcase.

In some commercial situations, there is a similar problem of loading a limited space with a selection of goods. Packaging for international freight often requires the packing of goods in standard sized containers whose volume is limited; if these containers are being transported by air, then their total weight is likely to be limited as well. Canal freight, which is much used on the mainland of

Europe, requires that goods be packed in the limited volume of a barge and within prescribed limits on the total weight of the cargo.

If a business is planning its spending programme on advertising, it must allocate funds between the media which are available. Some may be spent on radio and television commercials; some on newspaper and magazine advertisements; some on special offers. Another business may have a choice of research projects competing for equipment and salaries: some pieces of work may be more desirable than others, based on a prediction of the likely success; on the other hand, some may be more expensive than others.

Problems such as these can be readily examined by means of dynamic programming. However they differ from the routing problems we have looked at earlier. The general characteristic of the problems that we examined in the last chapter was that at each stage, a decision about a particular destination had to be made. Associated with this decision, there was some cost or benefit. In this chapter, we shall look at packing problems where the characteristic decision involves selection of a quantity of material. Once again, each decision we can make will have some cost or benefit linked with it. But the decision we take will also reduce the limited supply of some resource such as weight, money, or space. The decision may be simply whether or not to take some item; alternatively, it may concern the quantity of the resource that we take. Because these problems have come about from examples of loading, they are sometimes referred to as "knapsack problems".

One important aspect of this type of decision problem is that the use of dynamic programming assumes that the decisions are taken in sequence over time; at first sight, one might think that the choices were being made simultaneously, but it is not too unrealistic to take a different point of view and see them as being made one after the other ... "Shall I pack any of the first item?"... "Shall I pack any of the second item?" ... "Shall I pack any of the last item?" (and in each case, we may possibly add "And if I do, how many should I pack?").

3.2 A simple knapsack problem

Consider a situation where a vehicle has a capacity of 10 tonnes and there are 4 products which might be loaded on board; their weights and values are shown in Table 3.1. How should the vehicle be loaded in order to maximise the value of its cargo?

Table 3.1: Loading a vehicle of capacity 10 tonnes

Product number (i)	1	2	3	4
Weight (tonnes) W_i	2	3	4	6
Value V_i	8	12	15	19

(The problem could also be written as an integer programme in which the objective was to maximise the sum $8x_1 + 12x_2 + 15x_3 + 19x_4$ subject to two kinds of restriction, that $2x_1 + 3x_2 + 4x_3 + 6x_4 \leq 10$ and that x_1, x_2, x_3, x_4 all should be equal to 0 or 1.)

In a problem which is as small as this, it is possible to examine all the ways of deciding whether or not to take each product and quickly find the optimum. However, as in the last chapter, it is helpful to look at the way that dynamic programming will handle this type of problem.

The first thing we must do is to break the decision into a sequence of simple choices. The numbering of the products helps with this; the first decision will concern product number 1, the second will be about product number 2 and so on. At each step we will have to decide whether or not to include that product or not. For dynamic programming we need to define states and stages: the stages of the loading problem correspond to the number of decisions that have still to be made. At the start we shall have four decisions, then three, until all the decisions have been taken. The information that is sufficient to describe the state when we take these decisions will be the amount of space that remains in the vehicle. So there will be 10 tonnes available at the beginning; 10 tonnes or less after the first decision has been taken; and so on down to the last decision. (Although the person responsible for packing the vehicle will consider the items in the numerical order given, the dynamic programming formulation looks at them in the reverse order, since the stages are numbered to count down to zero.)

Suppose we are at stage n and there is capacity for T tonnes in the vehicle. If we define $f_n^*(T)$ as the value of the optimal load that we can prepare from this state and stage, we shall see that:

$$f_n^*(T) = \max\{f_{n-1}^*(T), V_n + f_{n-1}^*(T - W_n)\} \quad (\text{for } T \geq W_n) \quad (3.01)$$

$$= f_{n-1}^*(T) \quad (\text{for } T < W_n) \quad (3.02)$$

with

$$f_0^*(T) = 0 \qquad \forall T$$

In words, the best that we can do in stage n and with state T is the better of not loading any of product n ($f_{n-1}^*(T)$) and loading product n and facing the next decision with a state reduced by its weight ($f_{n-1}^*(T - W_n)$) provided that T is large enough to allow this to happen.

Quite clearly, $f_1^*(T)$ is only going to have two possible values: 0 or 19; it will be 0 if T is less than 6 and 19 if T is 6 or more. (In other words, when we reach the stage of having to make a decision about product number 4, we will only load it if there is space to do so, as measured by the state which we are then in.)

For $f_2^*(T)$ we need to use the recurrence relation. In theory, we could have many states T when we reach this stage and so we work out the values of $f_2^*(T)$

for all T from $T = 0$ to $T = 10$ (although we should note that with a little effort we might be able to eliminate several values of T as being impossible at stage 2). Then:

$$\begin{aligned} f_2^*(10) &= \max(f_1^*(10), 15 + f_1^*(6)) \\ &= \max(19, 15 + 19) = 34 \qquad (3.03) \\ f_2^*(9) &= \max(f_1^*(9), 15 + f_1^*(5)) \\ &= \max(19, 15 + 0) = 19 \qquad (3.04) \end{aligned}$$

...

$$\begin{aligned} f_2^*(1) &= f_1^*(1) = 0 \\ f_2^*(0) &= 0 \end{aligned}$$

This yields values for $f_1^*()$ and $f_2^*()$ in Table 3.2.

Table 3.2: Values for $f_1^*()$ and $f_2^*()$

T	0	1	2	3	4	5	6	7	8	9	10
$f_1^*(T)$	0	0	0	0	0	0	19	19	19	19	19
$f_2^*(T)$	0	0	0	0	15	15	19	19	19	19	34

As we move backwards through the decision process, we next calculate $f_3^*()$ using $f_2^*()$ (but **not** $f_1^*()$). Exactly the same logic as earlier allows us to calculate:

$$\begin{aligned} f_3^*(10) &= \max(f_2^*(10), 12 + f_2^*(7)) \\ &= \max(34, 12 + 19) = 34 \qquad (3.05) \\ f_3^*(9) &= \max(f_2^*(9), 12 + f_2^*(6)) \\ &= \max(19, 12 + 19) = 31 \qquad (3.06) \end{aligned}$$

...

$$\begin{aligned} f_3^*(1) &= f_2^*(1) = 0 \\ f_3^*(0) &= 0 \end{aligned}$$

Continuing in this way, we can find the values of $f_4^*()$ (using $f_3^*()$ but not $f_2^*()$ and $f_1^*()$) which are presented in Table 3.3. This table gives the value of all four functions, with T taking all values from 0 to 11 (and not 10, the capacity of the vessel). The extension is useful as a means of checking how sensitive the answer is to the capacity of the vessel. Given one extra unit of capacity, how much benefit would it give to the optimal loading? The answer is that it would

Table 3.3: Optimal values of the objective functions

T	0	1	2	3	4	5	6	7	8	9	10	11
$f_1^*(T)$	0	0	0	0	0	0	19	19	19	19	19	19
$f_2^*(T)$	0	0	0	0	15	15	19	19	19	19	34	34
$f_3^*(T)$	0	0	0	12	15	15	19	27	27	31	34	34
$f_4^*(T)$	0	0	8	12	15	20	23	27	27	35	35	39

increase the value by 4 units from 35 to 39. Later in the chapter we shall return to the subject of sensitivity analysis in more detail.

Table 3.3 is useful for answering questions about the value of the optimal policy; we know that it will be worth 35 units if we pack the vessel in an optimal way. But this function value doesn't directly tell us what the optimal set of decisions should be. To find these, we shall need some more information than is provided in the table.

If we go back to the recurrence relationships:

$$f_n^*(T) = \max\{f_{n-1}^*(T), V_n + f_{n-1}^*(T - W_n)\} \qquad (\text{for T} \geq \text{W}_\text{n}) \qquad (3.07)$$

$$= f_{n-1}^*(T) \qquad (\text{for T} < \text{W}_\text{n}) \qquad (3.08)$$

we can readily identify the optimal decision in each state at each stage as being 'Use product n' or 'Do not use product n'. (It is possible that the optimal decision is for either of these; this leads to two (or more) equally good solutions.) This decision can be stored alongside the policy value and then used to identify the optimal policy or policies when the problem has been completely solved. If we do this for our example we will have the table below in which are recorded the optimal number of items (0 or 1) to be packed to give the optimal policy that has been recorded above.

Using these figures (Table 3.4) and the recurrence relationship (with its known values of W_i and V_i) we can find the optimal policy; $f_4^*(10)$ arose from using one unit of item 1, and so it was calculated from $f_3^*(8)$ which in its turn arose from using one unit of item 2, so it was calculated from $f_2^*(5)$ and which had come from using one unit of item 3 and none of item 4. (It is worth noticing that $f_4^*(11)$ comes from a different policy.) It is generally worth keeping a record of the decisions as the function is calculated, although, as is evident from the discussion just given, it is comparatively easy to find the optimal decisions from the final table of values of the objective function. The decisions are indicated using boxed characters.

3.3 Extending the formulation

Our discussion so far has been concerned with a loading problem in which there

Table 3.4: The optimal policy

T	0	1	2	3	4	5	6	7	8	9	10	11
$i_1^*(T)$	0	0	0	0	0	0	1	1	1	1	1	1
$i_2^*(T)$	0	0	0	0	1	1	0	0	0	0	1	1
$i_3^*(T)$	0	0	0	1	0	0	0	1	1	1	0	0
$i_4^*(T)$	0	0	1	0	0	1	1	1 *or* 0	1 *or* 0	1	1	1

was a limit that there should never be more than one item of each product in the vessel. This restriction need not apply, and if it doesn't, then we shall have a variant of the loading problem. Many of the features will be the same as earlier, but the decisions will be more complex: instead of being binary (0 or 1), the number of items of product n loaded will only be restricted by the space available in the vessel, as measured by the ratio of the remaining capacity (T tonnes) divided by the weight (W_n tonnes) of it. So we have a recurrence relationship (equation 3.09).†

$$f_n^*(T) = \max_{0 \leq i_n \leq \lfloor \frac{T}{W_n} \rfloor} \{f_{n-1}^*(T - i_n W_n)\} \tag{3.09}$$

This new possibility will make little difference to the method of calculating the optimal policy; we will once again start with the best way of packing the vessel with one type of product, then proceed to finding the best way with two types, and so on. The table of results will generally be different, because there will be some states and stages from which the optimal policy is to pack more than one unit of product, and the optimal policy may be quite different from the one which was found with the restriction to a limit of one unit. If we look at the example that was considered in the chapter earlier, we shall find that the table of results is shown in Table 3.5.

Table 3.5: The effect of multiple items on $f^*()$

T	0	1	2	3	4	5	6	7	8	9	10	11
$f_1^*(T)$	0	0	0	0	0	0	19	19	19	19	19	19
$f_2^*(T)$	0	0	0	0	15	15	19	19	30	30	34	34
$f_3^*(T)$	0	0	0	12	15	15	24	27	30	36	39	42
$f_4^*(T)$	0	0	8	12	16	20	24	28	32	36	40	44

† The notation $\lfloor x \rfloor$ is used to indicate the largest integer n such that $n \leq x$

The figures in Table 3.5 are all no smaller than the figures we saw earlier, in Table 3.3. The value of the optimum policy has increased from 35 to 40, and many other figures have increased correspondingly. We should expect the change between the solution to the problem with binary decisions and that with discrete decisions to be positive, because the latter has fewer restrictions inherent in it. The optimum policy $f_4^*(10)$ corresponds to either 5 units of item 1 or 2 units of item 2 with 2 units of item 1.

3.4 Further extensions

The problems we have looked at so far could have been easily posed as integer programmes; the first would have been the problem:

$$\text{maximise} \sum_{n=1}^{4} V_n i_n \text{ subject to } \sum_{n=1}^{4} W_n i_n \leq C \text{ and } i_n = 0, 1 \ \forall\, n \qquad (3.10)$$

while the second was similar:

$$\text{maximise} \sum_{n=1}^{4} V_n i_n \text{ subject to } \sum_{n=1}^{4} W_n i_n \leq C \text{ and } i_n \text{ integer } \forall\, n \qquad (3.11)$$

Less easy to formulate as integer programmes, but very amenable to treatment as dynamic programmes, are problems in which the value of multiple items of a product is a nonlinear function of the number of items packed. The nonlinearity may be in either direction; the value of two items may be more than, or less than, twice the value of one. Knapsack problems in which there are departures from linearity are generally more accurate models of an underlying reality than the ones we have looked at so far. They may arise in problems of loading (for example there may be products where it is only necessary to pack one item, and the benefit of packing any extra is small or zero) or in problems of investment (for example a shop could choose to increase the size of its stock of a sales line; the extra number of goods may have only a small effect on the sales of the item). Similar types of problem arise in the selection of advertising media.

Such nonlinear problems can be expressed as mathematical programmes (rather than integer programmes) assuming that the constant value V_n per unit of product n is replaced by a function $V_n(i_n)$ of the number of items of that product which are loaded. Since we are still working in integers, it is generally simplest to represent this function as a table. The example in Table 3.6 shows such data. How should a vessel holding 10 tonnes be loaded optimally?

The procedure to be followed is essentially the same as the one we used when there was no limit on the values of the decision variables (the i_n). However, to

Table 3.6: A nonlinear objective for a 10 tonne vessel

Product number (n)	1	2	3	4
Weight (tonnes) W_n	2	3	4	6
Value $V_n(1)$	8	12	15	19
$V_n(2)$	12	20	28	
$V_n(3)$	14	26		
$V_n(4)$	16			
$V_n(5)$	18			

evaluate the alternatives, we must replace the product $V_n i_n$ by the function $V_n(i_n)$, to give the recurrence relationship:

$$f_n^*(T) = \max_{0 \leq i_n \leq \lfloor \frac{T}{W_n} \rfloor} \{V_n(i_n) + f_{n-1}^*(T - i_n W_n)\} \tag{3.12}$$

together with the initial conditions $f_0^*(T) = 0 \quad \forall T$

Proceeding with the calculations in the obvious way, we find that the values for the objective function $f_1^*(T)$ are identical to those given earlier (Tables 3.2, 3.3 and 3.5). We should not be surprised at this, because there is no possibility of loading the vessel with more than one item and the value of a single item has not been changed between Table 3.1 and Table 3.6. However, if we apply the recurrence relationship to obtain values of $f_2^*(T)$ we shall find that it differs from that seen earlier in Table 3.2. The values are given in Table 3.7.

Table 3.7: $f_1^*(T)$ and $f_2^*(T)$ with nonlinear V_n

T	0	1	2	3	4	5	6	7	8	9	10
$f_1^*(T)$	0	0	0	0	0	0	19	19	19	19	19
$f_2^*(T)$	0	0	0	0	15	15	19	19	28	28	34

Two values have changed from Table 3.2 to Table 3.7. Both $f_2^*(8)$ and $f_2^*(9)$ have increased from 19 (corresponding to packing one item of item number 4) to 28 (two items of item number 3). If we continue in this fashion the values of the objective function will be those shown in Table 3.8 (where, once again, the range of values of T has been extended to $T = 11$ to allow for sensitivity analysis.

3.5 Sensitivity analysis in knapsack problems

Although the optimal solution to the knapsack problems has generally been expressed as a numerical value, we have seen that this can be easily rewritten

Table 3.8: Objective function with nonlinearities

T	1	2	3	4	5	6	7	8	9	10	11
$f_1^*(T)$	0	0	0	0	0	19	19	19	19	19	19
$f_2^*(T)$	0	0	0	15	15	19	19	28	28	34	34
$f_3^*(T)$	0	0	12	15	15	20	27	28	31	35	40
$f_4^*(T)$	0	8	12	15	20	23	27	28	35	36	40

as an optimal policy. This optimal policy is a set of integer variables which are non-zero at the optimum. By analogy with linear and integer programming, it is sensible to think of these non-zero variables as being the optimal basis, although this is an expression which is not widely used in dynamic programming. We may be interested in the effect on this basis of changes in the data, and particularly in the parameters associated with the products being loaded. For instance, what happens to the optimal basis (and thus to the value of the optimal solution) when the weight of a product is reduced by one tonne? Or what happens when its value is increased by three cost units?

In many cases we need to solve the problem afresh to answer these questions. But it is possible to state some simple ground rules which may eliminate this need for a few cases. It is sensible to look at such rules, and thereby to look at the ways that we can perform sensitivity analysis in knapsack problems.

Suppose that we have been trying to load products into a vessel of limited capacity. Let us assume that there are N products identified as $1, \ldots, N$; assume the vessel has capacity C tonnes and the products have known weights W_i and values V_i. The optimal solution will have value $f_N^*(C)$. Corresponding to this value will be one or more optimal bases. If we look at the effect of changing the parameters for an arbitrary product i, we need to think of the three possible situations that involve i at this optimum. These three are:

1) i is in a unique basis, or is in all the optimal bases;
2) i is in one of the optimal bases, of which there are several, including some from which i is absent;
3) i is not basic in any optimal solution.

The possible changes to weights W_i and values V_i are increases and decreases; putting these changes together with the three situations given above we have twelve possible situations to consider in Tables 3.9 and 3.10. These consider the three cases just listed and the possible changes that might be made to the weight and the value of an item.

3.6 Adding one or more dimensions

In all the examples so far, the decision-making process has been constrained

Table 3.9: The effect of varying W_i

	W_i reduced	W_i increased
Case 1)	no change until the basis changes, then $f_N^*(C)$ increases	no change until the basis changes, then $f_N^*(C)$ decreases
Case 2)	no change until the basis changes, then $f_N^*(C)$ increases	no change at all
Case 3)	no change until either the basis changes, then $f_N^*(C)$ increases or W_i cannot be further changed	no change ever

Table 3.10: The effect of varying V_i

	V_i reduced	V_i increased
Case 1)	$f_N^*(C)$ decreases until the basis changes and then either remains constant or decreases at a slower rate	$f_N^*(C)$ increases and there may be a change of basis which makes its rate of increase become larger
Case 2)	no change in the value of $f_N^*(C)$	$f_N^*(C)$ increases and there may be a change of basis which makes its rate of increase become larger
Case 3)	no change ever	no change until the basis changes and i becomes part of the basis, then $f_N^*(C)$ increases

by one dimension, the weights of the item. In several practical examples, as the introductory discussion suggested, there will be restrictions for other causes such as the volume of the vessel being loaded. One might, for example, be concerned with the optimal loading of a container designed to hold a maximum weight of 40 tonnes and with a volume of 100 cubic metres. The specification of the goods to be loaded would require a description of the value per item (and possibly the functional form of the value of more than one item), the weight of each one and the space occupied by each. (One could further imagine that there might be nonlinearities in the weight and volume to give scope for packing several items together. We shall ignore such possibilities here.) A typical set of data might take the form of Table 3.11.

If one ignored the constraints on volume, we would be faced with a problem similar to the nonlinear table (Table 3.6). It could be solved to give an objective

Table 3.11: Loading a vessel restricted by weight (40 tonnes) and volume (100 cubic metres)

Product number (i)	1	2	3	4	5
Weight (tonnes) (W_i)	10	7	5	4	2
Volume (cubic metres) (C_i)	18	20	16	15	5
Value of 1 item ($V_i(1)$)	30	20	18	14	10
Value of 2 items ($V_i(2)$)	50	38	28	16	18
Value of 3 items ($V_i(3)$)	65	50	32	18	25
($V_i(4)$)	73	55	36	20	30
($V_i(5)$)	N.A.	60	36	22	30
($V_i(6)$)	N.A.	N.A.	36	24	30
($V_i(7)$)	N.A.	N.A.	36	26	30

function value 125, corresponding to the decisions: 3 of product 5, 1 of product 4, 1 of product 3, 2 of product 2 and 1 of product 1. However, the total volume of this mix of products would be 104, violating the limit that had been placed on the volume of the vessel.

Similarly, if the constraints on weight were to be set aside, we would have another nonlinear set of data with limits due to the volume of the products and the vessel. The solution to this problem is 122, corresponding to loading as follows: 3 of product 5, 1 of product 4, 1 of product 3, 0 of product 2 and 3 of product 1. However, the total weight of this mix of products would be 45, violating the limit that had been placed on the weight of the vessel.

It is worth noting that if the solutions that were found in the preceding paragraphs satisfied the constraints which had been temporarily ignored, then we should have found a solution to the two-dimensionally constrained problem. This would save a great deal of effort, as we shall see.

The dynamic programming approach to problems where there are two or more limits on the resources being used is essentially the same as for those cases where there is only one such resource. We define an objective function whose value is the result of following an optimal policy from a given stage onwards, starting at a particular state. Values of the function are found by considering a recurrence relationship involving the consequence of a decision in a single stage followed by the optimal policy from the resulting state. The principal difference is to be found in the description of the states of the system being modelled.

In all the examples we have considered so far, the state has been identified by a single value; in the shortest route models, the state was a node of the network; in the simple knapsack problems it was the amount of space or weight left in the vessel. When there are two or more resources to be considered, one has to describe the state as a vector, whose components are the amounts still to be used. As a result, there will be many more possible states to consider than there were when the state was identified more simply.

Nonetheless, the form of the recurrence relationship will be a familiar one. If we define the state at stage n in terms of the remaining volume (S_n) and remaining weight (T_n), then the value of an optimal policy will satisfy the recurrence:

$$f_n^*(S_n, T_n) = \max_{i_n \in P_n} \left(V_n(i_n) + f_{n-1}^*(S_n - i_n C_n, T_n - i_n W_n)\right) \tag{3.13}$$

$$f_0^*(S_0, T_0) = 0 \qquad \forall S_0, T_0 \tag{3.14}$$

$$\text{where } P_n = \left\{0, 1, 2, \ldots, \min\left(\left\lfloor \frac{S_n}{C_n} \right\rfloor, \left\lfloor \frac{T_n}{W_n} \right\rfloor\right)\right\} \tag{3.15}$$

At each stage, we shall need to calculate a value of f^* for each possible state; in the example of Table 3.11, there could be $41 \times 101 = 4141$ such states, when all possible values of the remaining weight and remaining volume have been considered (zero is included in each case). A large proportion of these might be impossible, but the effort of demonstrating that a state could not occur at a given stage is likely to be comparable to the effort needed to calculate the value of the objective function. Clearly, the solution to this recurrence relation requires computer assistance. Although it would be possible to present the five tables of optimal values of f_n^* in a score of pages in this book, the benefit of this to a reader would be negligible. Table 3.12 gives some of the optimal values of f_5^* showing that the value of an optimal load will be 120, and that the first decision will be to pack 2 items.

If there were more possible states in either or both of the dimensions of resources, then the amount of calculation would be further increased. The tedium of calculations when there are more than two dimensions is still greater. With four resources, each of which can take 20 values, there will be 160 000 states to consider at each stage. Adding constraints makes dynamic programming problems become beyond the reach of even the fastest computers; the recognition of this difficulty quickly led early students to refer to it as "The curse of dimensionality". It is an obstacle which arises in many formulations of dynamic programmes, where the accurate description of the state of a system is made using several distinct variables. There can be no automatic way of avoiding the problem. In some cases it may be possible to redefine the state using fewer dimensions, but this cannot be guaranteed. Four to five dimensions seems to be generally accepted as the limit in most problem formulations, though this can be increased if some of the ways of describing a state are simple binary ones.

When faced with a dynamic programme involving a large number of variables in the state description, it is generally advisable to see whether they are all

Table 3.12: Optimal values of f_5^* (and the corresponding decision)

weight volume	100	99	98	97	96	95	94	93
40	120(2)	120(2)	120(2)	120(2)	118(4)	118(4)	118(4)	118(4)
39	118(2)	118(2)	116(4)	116(4)	116(4)	116(4)	116(4)	114(4)
38	118(2)	118(2)	116(4)	116(4)	116(4)	116(4)	116(4)	113(3)
37	118(2)	118(2)	116(4)	116(4)	116(4)	116(4)	116(4)	112(4)
36	112(4)	112(4)	112(4)	112(4)	112(4)	112(4)	112(4)	112(4)
35	112(4)	112(4)	112(4)	112(4)	112(4)	112(4)	112(4)	112(4)
34	112(4)	112(4)	112(4)	112(4)	112(4)	112(4)	112(4)	112(4)
33	107(3)	107(3)	107(3)	107(3)	107(3)	107(3)	107(3)	107(3)
32	107(3)	107(3)	107(3)	107(3)	107(3)	107(3)	107(3)	107(3)
31	100(2)	100(2)	100(2)	100(2)	100(2)	100(2)	100(2)	100(2)
30	100(2)	100(2)	100(2)	100(2)	100(2)	100(2)	100(2)	100(2)

relevant, by solving the less complicated problems which exclude each one in turn. Only if all the solutions violate the constraint which has been set to one side should one proceed to solve the full problem.

3.7 Sensitivity analysis

The topic of sensitivity analysis applied to problems where the state is described by two or more variables has many similarities to the one-dimensional counterpart. If an item is in the basis at the optimum, then reduction in the amount of any resource that it requires will not immediately change the optimal value of the objective function or the policy. At a critical value, the basis will change and a new policy with a larger value to the objective will become optimal. This new policy may include further units of the same resource or be totally different. If the amount of resource required is increased, then there will be a second critical value which will herald a new optimal policy and an objective function that is no larger than the first one.

For variables that are not in the basis at the optimum, an increase in the amount of resource that they require will never have any effect on the optimal solution. A reduction in their resource requirements will not affect the solution until they can become basic; this may never happen with a problem with a multi-dimensional description of the state, even though it could when the state was indexed by a single variable.

3.8 Exercises

(1) A barge has to be loaded with up to 14 tonnes of goods. These are packed into crates, and it is only possible to have one product type in each crate. The value of a crate of each of the four products and the weight of a crate is shown in the Table 3.13; what is the optimum policy for loading the barge?

Table 3.13: Loading a 14 tonne barge (exercise 1)

Product i	1	2	3	4
Weight (tonnes) W_i	5	3	5	7
Value (£1000s) V_i	13	9	17	23

(2) Solve the following knapsack problem, expressed as an integer programme to be maximised:

$$\begin{aligned} \text{maximise} \quad & 7x_1 + 6x_2 + 2x_3 = Z \\ \text{subject to} \quad & 4x_1 + 3x_2 + x_3 \leq 10 \\ & x_1 \leq 3, x_2 \leq 1, x_3 \leq 2 \\ & x_i \text{ integer} \end{aligned}$$

(3) Consider a general knapsack problem expressed as:

$$\begin{aligned} \text{maximise} \quad & V_1x_1 + V_2x_2 + V_3x_3 + \ldots + V_nx_n = Z \\ \text{subject to} \quad & W_1x_1 + W_2x_2 + W_3x_3 + \ldots + W_nx_n \leq w \\ & x_i \text{ integer} \end{aligned}$$

and assume that the variables have been arranged so that they are in decreasing order of value for weight, that is:

$$\frac{V_1}{W_1} > \frac{V_2}{W_2} > \frac{V_3}{W_3} > \ldots > \frac{V_n}{W_n}$$

Show that the "best" item, number 1, will be used in the optimal policy provided that

$$w \geq \frac{V_1W_1}{V_1 - W_1(V_2/W_2)}$$

and not otherwise. (This result is sometimes referred to as a turnpike theorem, from the American usage of that word. It derives its name from the idea that for some road journeys, it is worthwhile using a route which is longer than the minimum if the speed of travel along the alternative will compensate for the extra mileage. This advantage only is noticed when the journey exceeds some minimum distance. Calling it a "motorway theorem" in the United Kingdom would not have quite the same ring to it! And with some U.K. motorways, the advantage would not be noticed!)

(4) Chagford airport has space for up to 6 extra flights on Saturdays in summer. An airline proposes to make use of these for holiday flights and makes estimates of the contribution to profits that would follow from different numbers of flights to each of three destinations. These profit contributions are shown in Table 3.14.

Table 3.14: Contributions to airline profits (exercise 4)

Number of flights	0	1	2	3	4	5	6
To Tenerife	0	250	450	600	700	750	775
To Majorca	0	300	525	675	750	775	775
To Alicante	0	225	425	600	750	875	975

What mixture of flights would maximise the airline's extra profits? How would this be affected if the airport could only fit in 5 extra flights? Or if there were to be space for 7 extra flights (assume that the company would not send all 7 to the same destination).

Table 3.15: Ms Black's stamps (exercise 5)

Item ↓ Expenditure	£0	£150	£300	£450	£600	£750
Kiloware	0	100	175	225	260	280
Old album leaves	0	120	210	250	260	270
Postal History	0	80	155	225	290	350
Classic G.B.	0	85	160	225	280	325

(5) Penelope Black is a stamp dealer who intends to invest £1200 in stock before attending a "stamp fair". The wholesaler who supplies her makes special offers for multiples of £150 being spent on any of four types of product, so Penelope always buys in such multiples. Based on her experience at stamp fairs, she expects to make the profits from differing expenditures on her material shown in Table 3.15. What should she buy?

(6) An engineer responsible for the computer network in Honiton University has discovered that there are four electronic components whose failure might mean that the network would require maintenance that would last for over an hour. He proposes that some of these components be given one or more back-up units which would increase the reliability of the network. The university is short of money and cannot afford more than £1000 at present. According

to accurate evidence of the reliability of the four components, Table 3.16 has been produced which shows the costs of back-up units and the probability that the component and all its back-ups will work reliably during a working month. These probabilities may be multiplied to give the probability that the network will work reliably for the month. Formulate the investment problem as a dynamic programme and find the optimal set of back-up units to install.

Table 3.16: Honiton University communications network (exercise 6)

Component	Amni	Emni	Imni	Omni
Cost per unit (in £s)	100	200	300	200
Probability: 0 back-ups	0.7	0.8	0.75	0.75
Probability: 1 back-up	0.91	0.96	0.9325	0.9325
Probability: 2 back-ups	0.973	0.992	0.9844	0.9844
Probability: 3 back-ups	0.9919	0.9984	0.9961	0.9961
Probability: 4 back-ups	0.9976	0.9997		0.9990

(7) Voters in the election to the Ruritanian senate were divided in their support of five parties. The voting figures are given in Table 3.17, where the expected number of seats in the senate are given; Ruritania uses proportional representation. Since fractions of seats cannot occur, it will be necessary to appoint representatives so that the maximum difference between the actual numbers of seats taken by each party and the expected number of seats is minimised, while retaining a senate of 15 members. Formulate the problem as a dynamic programme and find the best allocation. How would you test the sensitivity of the answer? Do you consider that this is a fair way of allocating the seats?

Table 3.17: The Ruritanian senate (exercise 7)

Party	Votes cast	Expected seats
Indigo	13351	1.5519
Blue	4732	0.5500
Green	58440	6.7928
Yellow	21967	2.5533
Orange	30559	3.5520
Total	129049	15

(8) There is an influenza epidemic in the state of Ruatha. Four towns have hospitals where local citizens can be immunised. However, there is only enough vaccine to treat 50 000 people, and (for various reasons) this must be assigned to the hospitals in units of 5 000 doses. Staff in the OR department at the Health Ministry have used their computer models of the spread of epidemics and have predicted the expected numbers of influenza patients in each hospital as a function of the number of people immunised there. These numbers are given in Table 3.18. Formulate the problem of assigning the vaccine to hospitals as a dynamic programme whose objective is to minimise the total number of influenza patients in the four hospitals.

Table 3.18: Expected numbers of influenza patients (exercise 8)

Doses of vaccine	Hospital 1	Hospital 2	Hospital 3	Hospital 4
0	50000	70000	75000	90000
5000	35000	50000	50000	65000
10000	15000	30000	35000	50000
15000	10000	15000	15000	30000
20000	8000	8000	5000	20000
25000	5000	3000	3000	10000

(9) The Information Technology (I.T.) department at Plymouth Coal plc is developing a new geographical database system. The newly appointed company chairman has told the department that she wants the project to be completed in six weeks' time, and is willing that up to six of the company's systems analysts should be allocated to the project in order to maximise the probability of successfully achieving this target. The director of the I.T. department has spent the morning estimating the probability of successfully completing each of the four parts of the database system by the deadline. All of these must work in the final product. These probabilities depend on the numbers of extra analysts and are detailed in Table 3.19. Formulate the assignment problem as a dynamic programme and solve it.

Table 3.19: Probability of success as a function of the number of systems analysts (exercise 9)

Analysts	Work	Video	Interface	Database	Software
0		0.5	0.5	0.6	0.6
1		0.75	0.7	0.8	0.75
2		0.8	0.9	0.85	0.85
3		0.95	0.9	0.9	0.85
4		0.95	0.9	0.95	0.85

4

Deterministic Production and Inventory Models

4.1 Introduction and a simple production problem

In many industries, only a small range of different products are made. Items are produced to order with order quantities known in advance; one of the problems for such industries is the scheduling of the amounts produced and the consequent inventories from planning period to planning period. Holding a stock of finished goods costs money; there is thus a cost which is a consequence of large production runs. Preparing the production line ("tooling up") and the loss of time and productivity while workers learn the production process again are two reasons why there is a cost associated with a short production run. The two costs must be balanced. There may also be storage and production constraints to be observed.

We shall consider a simple example based on such an industry. Imagine a workshop which makes fitted kitchen units to order. Because each product is "one-off", it isn't possible to hold a stock of finished goods to be sold "off the shelf". But some components are reasonably standard; the doors to cupboards and the drawers are examples. So it is possible to think of preparing them in advance of the time when they will be needed. Suppose that the workshop has orders which result in requirements for cupboard doors as in Table 4.1. The owner, at the start of the month of January, wants to plan his monthly schedule of production of these almost standard items to minimise his costs.

Tooling up for a production run costs £50 with an overhead of £12 per door; stockholding costs £2 for each door held from one month to the next. Orders must be satisfied. Assuming that the inventory level at the start of January is

Table 4.1: Requirements for a joinery business

Month (i)	January	February	March	April	May	June
Requirements (D_i)	15	12	20	32	25	12

zero, we know that there must be a production run during the month, to produce at least 15 doors. We shall assume that the aim is to end the month of June with a zero inventory as well, although this is an assumption that we can remove if desired.

Other problems arise in similar settings. A manager may have several pieces of work which need to be done in some sequence and which should satisfy some deadlines. If it is impossible to find a sequence which satisfies all the deadlines, which sequence is the best? The workshop that was introduced a little earlier may have orders for several kitchen units; some customers may have specified dates for delivery and may be unwilling to accept them being late, while others may accept late fulfilment of orders provided some discount is offered. A student may have several assignments to complete with differing due dates; in what order should he or she do the work? The second part of the chapter will look at such scheduling problems.

4.2 Formulating the problem of the doors

The problem for the workshop owner is obviously a sequential one; at the start of every month, he must make some decision about whether or not to tool up and make some more items. If he does make some more items, he needs to decide the number which will be constructed. The information he will have at the time will be a single figure, the number of items that he has in stock which have been made in earlier production periods and not yet used. We recognize this as the appropriate description of the state of the system, and we recognize that the sequence of monthly decisions corresponds to the set of stages in which the decisions are being made. To complete the formulation of the problem as a dynamic programme, we shall need to deduce a recurrence relationship and an appropriate objective function. The owner wants to minimise total cost, so the objective function will be based on a cost. The recurrence relation will link the costs in successive months. Let us define $f_n^*(I_n)$ as the cost of an optimal policy in month n when the inventory level is I_n at the time of making the decision. The workshop owner will decide a quantity to make, x items, and the cost will be $f_n(I_n, x)$ as a result. At the end of the month, the inventory level will be $I_n + x - D_n$; this will incur holding charges and will also be the inventory at the time of making the next decision, which will be referred to as month $n-1$ as we count backwards to the last month of the planning horizon. Instead of using the specific values of the inventory and production costs given earlier, we will assume that in the nth month the fixed cost of a production run is F_n and the variable cost is P_n per item; stockholding costs h_n per item that is kept from one month to the next. (So we ignore the costs of storing items during a month.) Using the principle of optimality, we can write

$$f_n(I_n, 0) = h_n(I_n - D_n) + f^*_{n-1}(I_n - D_n) \qquad (4.01)$$

and

$$f_n(I_n, x) = F_n + P_n x + h_n(I_n + x - D_n) + f^*_{n-1}(I_n + x - D_n) \qquad (4.02)$$

(if $x > 0$). Then

$$f^*_n(I_n) = \min_{x \geq D_n - I_n} (f_n(I_n, x)) \qquad (4.03)$$

4.3 Two simplifying theorems

At first sight, one could despair about the amount of work that this problem might entail. Do we really need to calculate f^*_n for all the possible values of the inventory at the beginning of the month? Fortunately, the answer is no. A very useful result proves that we only need to consider a limited number of possible decisions and hence inventory levels for each decision point. The result may be stated informally: *In an optimal policy, the only production quantities which need to be considered are those which ensure that the quantity in stock as a result of the production in that month will be sufficient to meet all the demand for a whole number of months and production will only take place if there is insufficient stock to meet the demand of the current month.* In consequence, it follows that with an inventory level of zero at the start of the planning period the only inventory levels which may occur in an optimal policy are:

$$\begin{aligned} I_n &= 0 \\ \text{or } &= D_n \\ \text{or } &= D_n + D_{n-1} \\ &\dots \\ \text{or } &= D_n + D_{n-1} + \dots + D_1 \end{aligned} \qquad (4.04)$$

If the opening inventory level is not zero, the result will still apply, but the consequence will not; there will be an initial period during which the inventory is used up. During that time, the inventory at the start of a month may take values other than the ones listed.

We may state this as a formal theorem thus:

In an optimal production plan:

a: the amount produced should be sufficient to meet the demand for a whole number of time periods;

b: there should only be production in a given time period if the opening inventory is not sufficient for the demand of that time period.

Proof: Suppose the optimal production plan is characterised by inventories $\{I_n^*\}$ and production quantities $\{x_n^*\}$.

a: Suppose that the optimal production plan does not yield production quantities which meet the demands for a whole number of time periods. Then there is some time period t, for which $0 < Y = I_t^* < D_t$ and $x_t^* > 0$, that is, the inventory at the start of the period is non-zero and not enough to satisfy the demand of that period. So one produces goods. Then the policy is not optimal, because money could be saved by changing the production plan. By increasing Y to $Y + 1$, there will be an increased charge $P_\tau + h_\tau + h_{\tau-1} + \ldots + h_{t+1} - P_t$ (where τ is the time period in which the items were produced which caused the violation of the assumption). If this is not equal to zero, the total cost of the so-called optimal policy can be reduced by varying Y within the limits that have been assumed, which contradicts the assumption that the policy was optimal. The limits on Y were assumed to be inequalities; to minimise the cost, Y should be equal to one or other of the extreme values. (This may further reduce the cost). If the change in the cost is zero, then there is no loss of generality in taking Y at its upper limit and this will mean that case b) applies and the policy is not optimal.

b) Suppose that in some month t, the opening inventory I_t^* is non-zero (and from what has just been shown, it must be $\geq D_t$) and there is some production $x_t^* > 0$. This cannot be optimal. Because items have been made in an earlier time period to be used in period t, the total cost per item ($=P_\tau + h_\tau + h_{\tau-1} + \ldots + h_{t+1}$ for some time period τ) is less than the cost per item made in period t, otherwise it would not be worthwhile having kept items and having a production run. So $P_t > P_\tau + h_\tau + h_{\tau-1} + \ldots + h_{t+1}$ for some τ. On the other hand, items which are made during time period t incur the same costs of being kept in stock $h_t + h_{t-1} + \ldots = FC$ (say) as those which are in the opening inventory, and so there will come a time period in the future when the former are preferred. But this will mean that $P_t + FC < P_\tau + h_\tau + h_{\tau-1} + \ldots + h_{t+1} + FC$ and so, subtracting the same term from both sides we will have the contradictory statement $P_t < P_\tau + h_\tau + h_{\tau-1} + \ldots + h_{t+1}$. Thus it will be impossible to have production in a period in which the opening inventory is large enough to meet the demands of that period.

4.4 Solving the problem of the kitchen doors

We can now return to the recurrence relations for the dynamic programme and simplify them. We had obtained:

$$f_n^*(I_n) = \min_{x \geq D_n - I_n} f_n(I_n, x) \tag{4.05}$$

where

$$f_n(I_n, 0) = h_n(I_n - D_n) + f^*_{n-1}(I_n - D_n) \quad (4.06)$$

$$f_n(I_n, x) = F_n + P_n x + h_n(I_n + x - D_n) + f^*_{n-1}(I_n + x - D_n) \quad (4.07)$$

(if $x > 0$)

The permissible values of x have been limited so that:

$$f^*_n(I_n) = h_n(I_n - D_n) + f^*_{n-1}(I_n - D_n) \quad (4.08)$$

if $I_n \geq D_n$ and

$$f^*_n(I_n) = \min_x \left(F_n + P_n x + h_n(I_n + x - D_n) + f^*_{n-1}(I_n + x - D_n)\right)$$

$$\text{where } x = \left(\sum_{\tau=t'}^{\tau=n} D_\tau\right) - I_n \quad \text{for some } t' \quad (4.09)$$

if $I_n < D_n$. Needless to say, the limited number of possible values of x to be considered in these expressions will simplify the calculation a great deal. If there is an upper limit on the size of the inventory as well, there may be still fewer.

In the discussion which follows, we shall assume that the opening inventory is zero; the alternative means a straightforward change to the analysis, which on first encounter could be confusing. As usual, the simplest approach is to work backwards from the final time period, 1 (June), calculating the optimal values of the objective function for the two possible opening inventories for this period. So we have:

$$f^*_1(0) = F_1 + P_1 D_1 \quad f^*_1(D_1) = 0 \quad (4.10)$$

Then there will be three possible opening inventories for time period 2 (May), and these yield:

$$f^*_2(0) = \min\Big(F_2 + P_2(D_2 + D_1) + h_2 D_1 + f^*_1(D_1),$$
$$F_2 + P_2 D_2 + f^*_1(0)\Big) \quad (4.11)$$

$$f^*_2(D_2) = f^*_1(0) \quad f^*_2(D_2 + D_1) = h_2 D_1 + f^*_1(D_1) \quad (4.12)$$

The problem is solved by working backwards to calculate $f^*_N(0)$. Using the data with which this section opened, this process can be illustrated: so we have $F_n = 50$, $P_n = 12$ and $h_n = 2$ $(\forall n = 1, \ldots, 6)$ with demands given in Table 4.2.

Table 4.2: Kitchen door orders

Month (i)	6	5	4	3	2	1
Requirements (D_i)	15	12	20	32	25	12

We then have:

$$f_1^*(0) = 50 + 12 \times 12 = 194$$
$$f_1^*(12) = 0$$

$$\begin{aligned} f_2^*(0) &= \min(50 + 12 \times (25 + 12) + 2 \times 12 + f_1^*(12), \\ &\qquad 50 + 12 \times 25 + f_1^*(0)) \\ &= \min(518, 544) = 518 \qquad \text{(Produce 37)} \\ f_2^*(25) &= f_1^*(0) = 194 \\ f_2^*(25 + 12) &= 2 \times 12 + f_1^*(12) = 24 \end{aligned}$$

Applying the recurrence relation for f_3^*, the values of the inventory to be considered are $\{0, 32, 57, 69\}$ and we find that:

$$f_3^*(0) = 952 \qquad \text{(Produce 32)}$$
$$f_3^*(32) = 518 \quad f_3^*(57) = 244 \quad f_3^*(69) = 98$$

Similarly for f_4^*, f_5^* and f_6^*:

$$f_4^*(0) = 1242 \qquad \text{(Produce 20)}$$
$$f_4^*(20) = 952 \quad f_4^*(52) = 582 \quad f_4^*(77) = 358 \quad f_4^*(89) = 236$$

$$f_5^*(0) = 1426 \qquad \text{(Produce 32)}$$
$$f_5^*(12) = 1242 \quad f_5^*(32) = 992 \quad f_5^*(64) = 686 \quad f_5^*(89) = 512 \quad f_5^*(101) = 414$$

$$f_6^*(0) = 1640 \qquad \text{(Produce 27)}$$

Thus we find that the optimal policy for the manufacturer will cost £1640. What policy will cost this sum? We have noted the source of each value f_n^* as we calculated it, but it is as easy to identify the policy at the end of the calculation. All we have to do is find which production quantity satisfies the recurrence relationship as an equality. Here our procedure is almost exactly the same as in the stagecoach problem when we wanted to identify the steps in the optimal path. So we know that:

$$1640 = 50 + 12x + 2(x - 15) + f_5^*(x - 15)$$

for some production quantity x; x is limited by the possible inventory levels allowed at stage 5 to be one of $\{15, 27, 47, 79\}$ and simple arithmetic shows that x should be equal to 27. We can then work forwards, knowing that the optimal policy is fixed by $f_5^*(12)$ which means that the production is zero in the second month and that the optimal policy is fixed by $f_4^*(0)$.

In this case of constant costs in each period, the optimal solution could also be recognized without using the dynamic programming formulation. The periods in which there has been production for more than one time period are those in which it is cheaper to hold stock than to set up a production run. Thus the optimal policy has only held goods from one period to the next when the demand for goods in the second period has been less than $\frac{F_{t-1}}{h_t} = \frac{50}{2} = 25$; if the costs were different between periods, then it would be more difficult to pin-point the periods in which there should be production by such a heuristic approach.

4.5 Some developments and extensions of the model

There are several ways in which we might extend this model and use the same approach. Essentially, at each stage of the sequence of decisions, there is a need to make a choice between production and no production. For each option, a cost is known, which includes factors for setting up production runs and storing surplus goods from one time period to later ones. We have made very few assumptions about the nature of this cost function, and so we could move away from the limited linear model which we used in the model for the kitchen unit supplier.

Typical cost functions that one might consider are:

1) those where the set-up and production costs depend on time; production at different times of year may cost different amounts; overall, the production costs are linear functions of the quantity produced.
2) those where the stockholding costs are time-dependent; for example, an item which is liable to perish may cost more to keep in warm weather than in cold.
3) those where the cost functions for production and/or stockholding are non-linear. Economies of scale may mean that large production runs are cheaper per item than small ones. (In a crude way, imposing a maximum level to the inventory is an example of a nonlinear cost function; the cost per item becomes infinite when the maximum is reached.)

Other possibilities with the simple model are to impose restrictions on the maximum and minimum rates of production if we were interested in a single type of item being produced. If we are looking at the production of several different items on the same equipment, then we might use a cost of changing the production item instead of the fixed charge seen earlier.

4.6 Sensitivity analysis of production planning models

The optimum policy we deduced for the scheduling of production and inventory could be expressed in several ways. In its simplest form, it could be written as a list of those months when there would (or would not) be production. Because of the theorem derived above, the quantities to be produced can be deduced from this information; hence the cost of the policy can be found. What happens to the policy when we change the parameters of the model? The parameters of interest are: fixed costs, variable inventory costs and order quantities for each time period. (We may ignore the possibility of variable production costs, since the only part of the solution that these will alter is the total cost; whenever production occurs, the cost per item of manufacture will be incurred.)

If we examine each of these in turn, we can make outline statements about the sensitivity of the solution to changes in their value. A reduction in fixed cost will make it more attractive to produce than to store goods; so the tendency will be to increase the frequency of production runs and reduce their size. Conversely, an increase in fixed cost will tend to concentrate the production into as few runs as possible. In each case there will be critical values for the parameter which determine when a policy changes. A reduction in the cost of holding stock from one time period to the next will have the same effect as an increase in the fixed production costs; more inventory will be held from one time period to the next, because it will be more attractive to do this than incur the costs of starting a production run. Similarly an increase in the cost of the inventory will tend to increase the number of time periods when there are production runs. Critical values for this cost will also exist. Changing the number of items needed in some or all of the time periods will affect the policy for two reasons. An increased demand in a month in which the optimal policy is to service the demand from storage will affect the policy when it becomes worthwhile to produce goods rather than use stored ones. In the same way, a reduction in the demand in a month when the optimal policy is to produce may tip the balance in favour of using stored items rather than starting a production run. The existence of a maximum level for the inventory may act as a restraint on production quantities in some months in the optimal policy; changing the demand figures may mean that this maximum is achieved after the demand for items in a different number of months has been satisfied.

4.7 Scheduling

In many manufacturing industries, managerial decision-making covers not only how much of a product is to be made, but also when it is made and in what sequence several products will be prepared. This sort of consideration becomes of especial importance when there are several products which compete for limited space or resources within a factory environment. "Which one should be made first?" is the sort of question that schedulers have to worry about. Additionally, there may be some targets by which time the work should be completed. In this section, we shall be looking at a simple scheduling problem, where a number of different items have to be made and it is desirable that each piece of work be completed on or before certain dates. An everyday example would be the garage where several car-owners leave their vehicles at the start of the day and arrange times when they wish to collect them later on or the problem of a student with several assignments that was mentioned in the introduction to this chapter.

In the jargon of scheduling, problems such as these are often referred to as being job shop problems. A **job shop** consists of a set of general purpose machines which perform operations on items which are being processed; we call these jobs. (There is no reason to limit the study to machines; the concepts are equally applicable to situations where staff have specific skills and abilities, and a piece of work requires attention by some or all of them in a particular order.) A job may well be a unique customer order, or it may be an item which is in regular production. The most obvious type of job shop is one where items are manufactured, but the same basic arrangement applies to many other environments such as maintenance and construction. We shall refer to the problem in terms of manufacturing and machines in this section.

The job shop scheduling problem consists of determining the best order for processing a set of jobs through the machines. Here, as throughout operational research, we need to determine precisely what we mean by the expression "best". There are several possible objectives in scheduling and these correspond to different priorities for the scheduler. Within this overall description, there are numerous possible problems. In this respect, scheduling is very similar to queuing.

If the machines of interest are all idle to start with, and all the jobs are ready at the same time to be processed, then the problem is said to be **static**. If the jobs arrive randomly, or are spread out with arrival times across a period of time, then the problem is **dynamic**.

Numerous objectives have been suggested for static scheduling problems. The most common is that of minimising the time when all jobs have been completed (the **makespan**). Alternatively, if jobs have times when they are due for completion, we may measure how late they are when they are completed and minimise the average **lateness** of a set of jobs. More realistically, instead of measuring the lateness (defined as the difference between the completion time and due time, which may thus be negative) we measure **tardiness** which is only of interest if the job is completed after its due time. It is sometimes desirable to weight some of the jobs as being more significant than others, so we would want to minimise a weighted sum of performances. The average **flow time** is of concern to some schedulers; this is the time that a job spends being worked on, and is of particular interest in dynamic scheduling problems.

4.8 Dynamic programming methods for scheduling problems

Dynamic programming is useful for solving some types of scheduling problems. The idea behind the use of DP in scheduling is that the optimal schedule for a set of jobs is made up of the optimal schedule for a subset of jobs followed by the optimal schedule for the complement of that subset given the later starting time(s) that results from having done some of the jobs. And thus we have the seeds of a DP formulation. We shall look at a simple example, of finding the best schedule for five jobs on one machine with the objective of minimising the mean tardiness (or the total tardiness). (More general problems of scheduling are covered by French (1982) and Johnson & Montgomery (1974), amongst other textbooks.) Suppose that our jobs have deadlines as shown in Table 4.3.

Table 4.3: Processing times and deadlines for jobs

Job i	1	2	3	4	5
Processing time (p_i)	24	8	24	31	18
Due date (d_i)	53	29	39	50	69

To consider this problem using dynamic programming, we need to define the essential features of any such formulation. Therefore we define the states of the system, a series of stages, an objective function and a recurrence relation which connects the successive states and stages. The stages are comparatively obvious. The person responsible for devising the schedule has to decide which jobs to schedule next, and the decisions are effectively taken as each job finishes. Therefore, there are five stages, corresponding to needing to schedule one job, two jobs, three jobs, four jobs and five jobs. At each stage there will be a number of possible states, corresponding to the possible sets of jobs which have been completed by that stage. The decision which answers the question "What next?" follows from considering the list of jobs that are waiting to be scheduled. The objective of the model is to minimise the mean tardiness, and as this is simply a multiple of the total tardiness, it is a straightforward transformation to the problem to take minimising the total as being the objective. When you decide on a job to be scheduled, this decision fixes the tardiness for that job. The set of jobs which have been finished determines the time when the decision is being taken, the duration of the job being scheduled determines when it will finish, and the due date for the job will determine the contribution of that job to the total tardiness function. And thus it is possible to write down a recurrence relationship.

Suppose that the set of jobs to be scheduled is a set S, and that we are considering the decision once all the jobs in a set $J \subset S$ have been completed. The time will be $T = \sum_{j \in J} p_j$. The stage will be $n = |S \setminus J|$. Then the total tardiness which follows from optimally scheduling the jobs in the complement of J, $S \setminus J$, will be:

$$f_n^*(J) = \min_{i \in S \setminus J} \left(\max(0, T + p_i - d_i) + f_{n-1}^*(J \cup \{i\}) \right) \tag{4.13}$$

Here the tardiness gives the single stage "cost" and this is followed by the optimal policy for the set of jobs that will have been scheduled at the time when the next decision is made. Working like this, it is useful to record the time (T)

as well as the set (J) for ease of calculation; so in the calculation that follows, the left hand sides of the equations are in the form $f_n^*(J)^{\equiv T}$:

$$f_1^*(\{1234\})^{\equiv 87} = \min(36 + f_0^*(\{12345\}) = 36 + 0) = 36$$

$$f_1^*(\{1235\})^{\equiv 74} = \min(55 + f_0^*(\{12345\}) = 55 + 0) = 55$$

$$f_1^*(\{1245\})^{\equiv 81} = \min(66 + f_0^*(\{12345\}) = 66 + 0) = 66$$

$$f_1^*(\{1345\})^{\equiv 97} = \min(76 + f_0^*(\{12345\}) = 76 + 0) = 76$$

$$f_1^*(\{2345\})^{\equiv 81} = \min(52 + f_0^*(\{12345\}) = 52 + 0) = 52$$

$$\begin{aligned} f_2^*(\{123\})^{\equiv 56} = \min(37 &+ f_1^*(\{1234\}) = 37 + 36 \\ 5 &+ f_1^*(\{1235\}) = 5 + 55) = 60 \end{aligned}$$

$$\begin{aligned} f_2^*(\{124\})^{\equiv 63} = \min(48 &+ f_1^*(\{1234\}) = 48 + 36 \\ 12 &+ f_1^*(\{1245\}) = 12 + 66) = 78 \end{aligned}$$

$$\begin{aligned} f_2^*(\{125\})^{\equiv 50} = \min(35 &+ f_1^*(\{1235\}) = 35 + 55 \\ 31 &+ f_1^*(\{1245\}) = 31 + 66) = 90 \end{aligned}$$

$$\begin{aligned} f_2^*(\{134\})^{\equiv 79} = \min(58 &+ f_1^*(\{1234\}) = 58 + 36 \\ 28 &+ f_1^*(\{1345\}) = 28 + 76) = 94 \end{aligned}$$

$$\begin{aligned} f_2^*(\{135\})^{\equiv 66} = \min(45 &+ f_1^*(\{1235\}) = 45 + 55 \\ 47 &+ f_1^*(\{1345\}) = 47 + 76) = 100 \end{aligned}$$

$$\begin{aligned} f_2^*(\{145\})^{\equiv 73} = \min(52 &+ f_1^*(\{1245\}) = 52 + 66 \\ 58 &+ f_1^*(\{1345\}) = 58 + 76) = 118 \end{aligned}$$

$$\begin{aligned} f_2^*(\{234\})^{\equiv 63} = \min(34 &+ f_1^*(\{1234\}) = 34 + 36 \\ 12 &+ f_1^*(\{2345\}) = 12 + 52) = 64 \end{aligned}$$

$$\begin{aligned} f_2^*(\{235\})^{\equiv 50} = \min(21 &+ f_1^*(\{1235\}) = 21 + 55 \\ 31 &+ f_1^*(\{2345\}) = 31 + 52) = 76 \end{aligned}$$

$$\begin{aligned} f_2^*(\{245\})^{\equiv 57} = \min(28 &+ f_1^*(\{1245\}) = 28 + 66 \\ 42 &+ f_1^*(\{2345\}) = 42 + 52) = 94 \end{aligned}$$

$$\begin{aligned} f_2^*(\{345\})^{\equiv 73} = \min(44 &+ f_1^*(\{1345\}) = 44 + 76 \\ 52 &+ f_1^*(\{2345\}) = 52 + 52) = 104 \end{aligned}$$

$$\begin{aligned} f_3^*(\{12\})^{\equiv 32} = \min(17 &+ f_2^*(\{123\}) = 17 + 60 \\ 13 &+ f_2^*(\{124\}) = 13 + 78 \\ 0 &+ f_2^*(\{125\}) = 0 + 90) = 77 \end{aligned}$$

$$f_3^*(\{13\})^{\equiv 48} = \min(27 + f_2^*(\{123\}) = 27 + 60$$
$$29 + f_2^*(\{134\}) = 29 + 94$$
$$0 + f_2^*(\{135\}) = 0 + 100) = 87$$

$$f_3^*(\{14\})^{\equiv 55} = \min(34 + f_2^*(\{124\}) = 34 + 78$$
$$40 + f_2^*(\{134\}) = 40 + 94$$
$$4 + f_2^*(\{145\}) = 4 + 118) = 112$$

$$f_3^*(\{15\})^{\equiv 42} = \min(21 + f_2^*(\{125\}) = 21 + 90$$
$$27 + f_2^*(\{135\}) = 27 + 100$$
$$23 + f_2^*(\{145\}) = 23 + 118) = 111$$

$$f_3^*(\{23\})^{\equiv 32} = \min(3 + f_2^*(\{123\}) = 3 + 60$$
$$13 + f_2^*(\{234\}) = 13 + 64$$
$$0 + f_2^*(\{235\}) = 0 + 76) = 63$$

$$f_3^*(\{24\})^{\equiv 39} = \min(10 + f_2^*(\{124\}) = 10 + 78$$
$$24 + f_2^*(\{234\}) = 24 + 64$$
$$0 + f_2^*(\{245\}) = 0 + 94) = 88$$

$$f_3^*(\{25\})^{\equiv 26} = \min(0 + f_2^*(\{125\}) = 0 + 90$$
$$11 + f_2^*(\{235\}) = 11 + 76$$
$$7 + f_2^*(\{245\}) = 7 + 94) = 87$$

$$f_3^*(\{34\})^{\equiv 55} = \min(26 + f_2^*(\{134\}) = 26 + 94$$
$$34 + f_2^*(\{234\}) = 34 + 64$$
$$4 + f_2^*(\{345\}) = 4 + 104) = 98$$

$$f_3^*(\{35\})^{\equiv 42} = \min(13 + f_2^*(\{135\}) = 13 + 100$$
$$21 + f_2^*(\{235\}) = 21 + 76$$
$$23 + f_2^*(\{345\}) = 23 + 104) = 97$$

$$f_3^*(\{45\})^{\equiv 49} = \min(20 + f_2^*(\{145\}) = 20 + 118$$
$$28 + f_2^*(\{245\}) = 28 + 94$$
$$34 + f_2^*(\{345\}) = 34 + 104) = 122$$

$$f_4^*(\{1\})^{\equiv 24} = \min(3 + f_3^*(\{12\}) = 3 + 77$$
$$9 + f_3^*(\{13\}) = 9 + 87$$
$$5 + f_3^*(\{14\}) = 5 + 112$$
$$0 + f_3^*(\{15\}) = 0 + 111) = 80$$

$$f_4^*(\{2\})^{\equiv 8} = \min(0 + f_3^*(\{12\}) = 0 + 77$$
$$0 + f_3^*(\{23\}) = 0 + 63$$
$$0 + f_3^*(\{24\}) = 0 + 88$$
$$0 + f_3^*(\{25\}) = 0 + 87) = 63$$

$$f_4^*(\{3\})^{\equiv 24} = \min(0 + f_3^*(\{13\}) = 0 + 87$$
$$3 + f_3^*(\{23\}) = 3 + 63$$
$$5 + f_3^*(\{34\}) = 5 + 98$$
$$0 + f_3^*(\{35\}) = 0 + 97) = 66$$

$$f_4^*(\{4\})^{\equiv 31} = \min(2 + f_3^*(\{14\}) = 2 + 112$$
$$10 + f_3^*(\{24\}) = 10 + 88$$
$$16 + f_3^*(\{34\}) = 16 + 98$$
$$0 + f_3^*(\{45\}) = 0 + 122) = 98$$

$$f_4^*(\{5\})^{\equiv 18} = \min(0 + f_3^*(\{15\}) = 0 + 111$$
$$0 + f_3^*(\{25\}) = 0 + 87$$
$$3 + f_3^*(\{35\}) = 3 + 97$$
$$0 + f_3^*(\{45\}) = 0 + 122) = 87$$

$$f_5^*(\{\})^{\equiv 0} = \min(0 + f_4^*(\{1\}) = 0 + 80$$
$$0 + f_4^*(\{2\}) = 0 + 63$$
$$0 + f_4^*(\{3\}) = 0 + 66$$
$$0 + f_4^*(\{4\}) = 0 + 98$$
$$0 + f_4^*(\{5\}) = 0 + 87) = 63$$

Thus the optimal sequence is $2 \to 3 \to 1 \to 5 \to 4$

The same approach can be used for other objective functions and for larger problems. The principal advantage of the method is that it saves a great deal of effort compared with complete enumeration. Thus, for our five job problem, we have had to look at $5 \times 1 + 10 \times 2 + 10 \times 3 + 5 \times 4 + 1 \times 5 = 80$ partial schedules compared with $5! = 120$ complete schedules using exhaustive enumeration. The advantages are more pronounced when you get larger problems. If you were dealing with a ten machine problem, the dynamic programming algorithm would require

$$10 \times 1 + 45 \times 2 + 120 \times 3 + 210 \times 4 + 252 \times 5$$
$$+210 \times 6 + 120 \times 7 + 45 \times 8 + 10 \times 9 + 1 \times 10 = 5120$$

partial schedules compared with the $10! = 3628800$ complete schedules of total enumeration.

To be precise, the dynamic programming formulation requires $n.2^{n-1}$ partial schedules; since $n!$ is approximately $\sqrt{2n\pi}\left(\frac{n}{e}\right)^n$ (Stirling's formula) the advantage should be apparent. The dynamic programming solution has complexity of the order of 2^n while total enumeration has complexity of the order of n^n.

The same general advantage will always be found and there are obviously ways that we could speed up the calculations by some sorting beforehand. But we have made a start on a problem with good results. Even so, the dynamic programming formulation won't be able to cope with problems which are much bigger. For the record, running a program on a personal computer to solve the scheduling problem took the times shown in Table 4.4. (Randomly generated data were used to create the problems.) The variation in time as the number of jobs increases is a consequence of two factors. There is the need to store more data and perform more calculations, and the larger problems require more searching through the stored f^* values.

Table 4.4: The time to solve scheduling problems of different sizes by dynamic programming

number of jobs	6	7	8	9	10
time (secs)	1.3	3.0	9.0	31.3	120.1

Sensitivity analysis in this particular formulation seems to have been totally ignored in the literature. Using dynamic programming has meant that we obtain a performance measure and the corresponding optimal sequence; but we have no idea of how good this is in comparison with another sequence. We know the best sequence which **starts** with a different job, but this may not be the second-best sequence. However, it does give an indication of roughly how much we could change one of the parameters in the optimal sequence by before there was a change. But the whole area of sensitivity analysis is wide open to further study.

4.9 The travelling salesman problem

One of the best-known operational research problems is that known as the travelling salesman problem (or salesperson). It is very easy to explain the problem, but exact solution methods are limited. One is given details of the distances (or times) for travel between each pair of cities in a set of N cities. The traveller starts in one of these, and makes a journey which visits each of the other $N-1$ once and once only and then returns to the starting point. The problem is to find the route which will minimise the total distance travelled. There is an extensive research literature on both the practical use of solutions and the development of algorithms for obtaining optimal (or near-optimal) solutions.

In the context of dynamic programming, there is a clear set of sequential decisions to be made. In the first city, the traveller must choose which will be the second one. In the second, he (or she) has to choose which will be the third, without returning to the first. In the nth, the choice is which will be the next one, and it must be different from all those that have already been visited. When the traveller reaches the Nth city of the journey, he has to return to the starting point, and so complete the tour. We can identify the stages by counting the decisions that remain; we can identify the decisions as the next city to visit. Optimality is determined as the tour with the least total length. What about the states? And what about the recurrence relation?

When we looked at problems of shortest routes, in chapter 2, the state was identified as the location of the traveller when a decision had to be made. That will not be sufficient for the travelling salesman, since there is the restriction of not returning to a city until the completion of the tour. So the state will be defined in two parts: (1) the location of the salesman; (2) either the set of cities that have been visited, or the set of cities that have not been visited. (These sets need not be ordered, since it is enough to know whether a journey to a chosen city is or is not permissible.) Hence there is a parallel with the scheduling problems just considered.

The recurrence relationship follows from this consideration of the states. In city i, at stage n, the salesman selects a city j to travel to, and the distance is taken from the data of the problem. From city j (at stage $n-1$) the optimal route through the remaining cities must be followed ending at the traveller's starting point. The relationship looks like that for a shortest route problem, with the extra detail of the set of cities that have been visited (remain to be visited). Defining city 1 as the starting point, and SV_n as the set of cities that have been visited up to and including stage n, we can write:

$$f_n^*(i, SV_n) = \min_{j \notin SV_n} \left(d_{ij} + f_{n-1}^*(j, SV_n \cup \{j\})\right)$$

with $f_0^*(i, \{1, 2, \ldots, N\}) = d_{i1}$; the answer will be $f_N^*(1, \{1\})$. The computation of this is feasible for small values of N, but the "curse of dimensionality" means that it becomes excessively slow as N increases. (Where the scheduling problem involved one state for each possible set of scheduled jobs, the travelling salesman problem involves several.) Exercise 11 is a simple example involving this formulation.

(The problem is generally stated in terms of distance, which implies that the "cities" can be located on a two-dimensional surface with symmetric distances, this need not be the case. Several applications generate "distance" matrices which are not symmetric, and whose "locations" cannot be mapped onto a plane. The dynamic programming approach is suited to any kind of matrix.)

4.10 Exercises

(1) Budleigh Beach Buggies (BBB) is a company which builds "off-road" cars to order. The company is part of a large organisation, so that staff can be moved between work for BBB and work involving other lines of business. Orders have been obtained for the first six months of next year, and these are presented in Table 4.5. If there is any production in a month, then it will cost £3000 to set up, and a fixed sum per car produced. Inventory charges are £250 per month

Table 4.5: Orders for beach buggies (exercise 1)

Month (i)	January	February	March	April	May	June
Requirements (D_i)	11	15	19	24	28	11

for the first ten cars, then £200 per month for each car in excess of ten. What is the optimal production schedule?
(2) In what ways is the solution to exercise (1) sensitive to the difference between the two inventory charges?

(3) How would the solution to exercise (1) be changed if the inventory cost per car were increased by £50 for January-February, by £40 for February-March, by £30 for March-April and so on?

(4) Consider the problem of a boatbuilder who has orders for dinghies as presented in Table 4.6.

Table 4.6: Demand data for dinghies (exercise 4)

Month (i)	January	February	March	April	May	June
Requirements (D_i)	5	3	6	4	2	1

A maximum of 10 dinghies can be constructed in a month, and it is necessary to produce at least one each month. The cost of construction per craft is a nonlinear function which may be summarised in Table 4.7. Storage costs £15 per month per dinghy: formulate the problem as a dynamic programme and deduce the optimal schedule.

Table 4.7: Construction costs for dinghies (exercise 4)

Number produced (j)	1	2	3	4	5
Cost per dinghy (C_j)	300	265	250	240	232
Number produced (j)	6	7	8	9	10
Cost per dinghy (C_j)	225	218	212	206	200

(5) A production manager has orders for widgets for the next five months as shown in Table 4.8. Production may be made in regular time at the rates shown per item or in overtime with the increased rates shown in the table. There are limits on the capacity of the plant for production which ensure that there should

be some production in each month; the limits on regular time production of widgets and on overtime production are shown. The holding cost per widget per month is £3. Formulate the problem of scheduling production at minimal total cost as a dynamic programme and find the solution.

Table 4.8: Orders and production details for widgets (exercise 5)

Month	Orders (items)	Regular time rate/item	Overtime rate/item	Regular time limit	Overtime limit
1	300	15	20	350	150
2	400	15	20	350	100
3	350	18	25	250	150
4	400	18	24	300	150
5	300	16	22	400	120

(6) A company manufactures thrimbles using one of two production processes. Each one requires the input of liquid squink and some of this may be recycled after use. In the first production process, one litre of squink is needed for every 8 thrimbles made, and 0.5 litres can be recovered afterwards. In the second production process, one litre of squink is needed for every 5 thrimbles made, and 0.75 litres can be recovered. Each production process requires one working day per run. Squink that has been used on seven successive days must be discarded, so the company aims to maximise the production that can be achieved from each batch of the liquid. Formulate this maximisation as a dynamic programme and find which machines should be used on each day.

(7) Consider the problem posed in exercise (6). Suppose that the processes yield $Y_1, Y_2 (< Y_1)$ thrimbles per litre respectively and that fractions r_1, r_2 are recycled. Deduce a general rule that determines which of the production processes should be used on each day of production.

(8) Five jobs have to be processed through a machine in a sequence that minimises the total tardiness. The processing times (p_i minutes) and the due dates (d_i minutes after processing starts) are given in Table 4.9. What is the optimal sequence?

(9) In exercise (8), consider the sensitivity of the solution to changes in: a) p_1; b) d_1.

(10) In exercise (8), the machine being used requires time c_{ij} minutes to change from job i to job j. How can the dynamic programme be amended to include this time? If $c_{ij} = 3|i - j|$ and the other data are not changed, what is the optimal sequence?

Table 4.9: Processing times and due dates (exercise 8)

Job J_i	1	2	3	4	5
p_i	75	81	51	34	23
d_i	108	174	140	153	45

(11) Table 4.10 gives the distance matrix for six locations. Select any five of these and solve the corresponding travelling salesman problem using a dynamic programming formulation; alternatively, write a computer program to solve the travelling salesman problem involving all six locations using a dynamic programming formulation.

Table 4.10: Travelling salesman distances (exercise 11)

	1	2	3	4	5	6
1	0	47	56	48	33	43
2	47	0	43	30	20	44
3	56	43	0	24	53	53
4	48	30	24	0	46	30
5	33	20	53	46	0	47
6	43	44	53	30	47	0

5

Optimisation of Functions

5.1 Introduction

So far we have examined a series of problems where the objective has been to minimise (or maximise) some function which has arisen in the context of a problem which could (reasonably) be expressed realistically as a sequential one, with variables which take one of a (small) set of discrete values. Dynamic programming has a further area of applicability when it is used for the optimisation of more general nonlinear functions of continuous or discrete variables subject to constraints. There is a very large range of methods for nonlinear optimisation in general, based on search algorithms or using the Karush-Kuhn-Tucker conditions for optimality (Karush (1939) and Kuhn & Tucker (1951)). The use of dynamic programming in nonlinear optimisation is limited to some fairly straightforward examples, but its use in these is both valuable and instructive.

5.2 One-dimensional minimisation

Many routines for optimisation of a nonlinear function rely on a series of searches along straight lines. There are several rules which determine which lines should be searched, depending on the amount of information one possesses about the function, its derivatives and its discontinuities. Each one relies on an efficient method of calculating the location of the minimum of the function along the straight line, or search line. Therefore there is considerable interest in finding reliable methods which locate the minimum of a function of one variable (which could be the parameter of a line in n dimensions) as rapidly (measured in terms of the number of function calculations that are needed) and as accurately as possible. These can then be incorporated in the search algorithm in n dimensions.

In the simplest problem, we want to minimise a function $g(x)$ of a scalar quantity x. The minimum of this function is at a point x^* which is one of the two desired outputs from the search. The other is $g(x^*)$. We have rules which allow us either to calculate this function at any point x or to compare the values of the function at two separate points x_1, x_2. Over the past thirty to forty years the search algorithms that have practical use for this have fallen into two categories. Some methods assume that $g(x)$ can be approximated by a polynomial function $g'(x)$ and that the location of the minimum of $g'(x)$ is a sensible approximation to the location of the minimum of $g(x)$. Successive approximations to $g(x)$, by a series of such polynomials $(g_1'(x), g_2'(x), \ldots)$ should yield a point estimate of the location of the minimum of the function. A second group of methods do not make such an assumption. They start with an interval in which it is assumed that x^* can be found, and by a succession of calculations of $g(x)$, reduce the interval to be as small as possible. We shall look at the way that dynamic programming can be used to define the points at which the function $g(x)$ should be calculated in an optimal plan of search in this latter group.

It is desirable to assume that the function has only one minimum in the interval; as a result, it is possible to deduce some information about the location of this minimum by comparing the value of $g(x)$ at two points within the interval. If the interval is (a, b) and the function is calculated at two points x_1, x_2 with $a < x_1 < x_2 < b$, then comparison of $g(x_1)$ and $g(x_2)$ must yield that either (i) $g(x_1) < g(x_2)$ or (ii) $g(x_1) \geq g(x_2)$. In case (i), the minimum of the function must lie in the interval (a, x_2); in case (ii), the minimum must lie in the interval (x_1, b). Figure 5.1 illustrates the two ways that the second condition can occur, with the minimum of $g(x)$ either in the interval (x_1, x_2) or the interval (x_2, b).

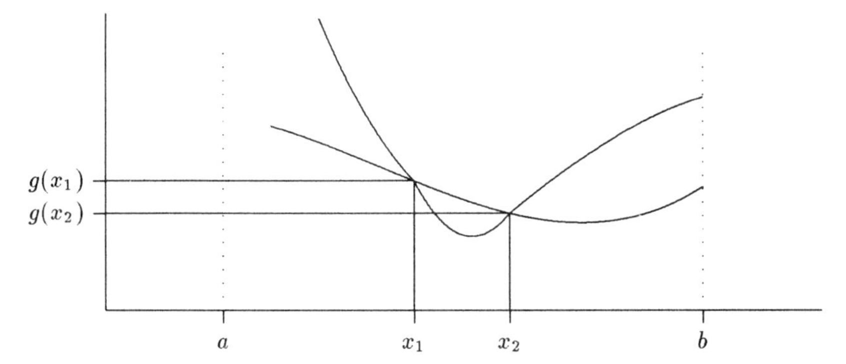

Figure 5.1: Information gained from two function values

(There is an unfortunate lack of symmetry between the two possible cases; (i) has a strict inequality, (ii) has the possibility of the function values being equal. Although this appears to be a problem, in most practical cases it is not. It would be possible to consider three outcomes of the comparison of $g(x_1)$ and $g(x_2)$ with two strict inequalities and equality as the options, but this would lead to a situation which would have zero probability – that two real numbers, calculated by a computer, were exactly equal. If the two function values were exactly equal, then we would know that the minimum was in (x_1, x_2), which is in both the intervals that cases (i) and (ii) provided.)

A further consequence of the cases (i) and (ii) above is that we know the value of $g(x)$ at some point within the interval already, which is useful for a subsequent repetition of the process which reduces the interval. Thus we ought only to need to calculate $g(x)$ at one further point on the next repetition or iteration of the method. By suitable choice of the interior points in the sequence, the search can reduce the length of the interval in an optimal fashion with every function calculation. The idea of such a succession of choices makes this problem a candidate for dynamic programming.

It is natural to suppose that the objective might be to seek to reduce the length of the interval to the smallest fraction of its original length possible given that we calculate $g(x)$ some fixed number N times. There often is a cost associated with each such evaluation of the function, and one would wish to gain as much information as possible for a fixed total cost. (Such costs may be due to the computer time involved, or the need to set up industrial experiments involving resources of time, materials and manpower.) So, if one started with an interval of length l_0 and an interval (a, b), with $b = a + l_0$, the length after N function evaluations might be l_N and we would want to minimise $\frac{l_N}{l_0}$.

We could solve this directly for small values of N. When $N = 1$, there is no advantage gained by calculating the function so $l_1 = l_0$; when $N = 2$ then the minimum value of l_N is found by choosing the two points x_1, x_2 in the interval so as to minimise the maximum of

$$\frac{x_2 - a}{b - a} \quad \text{and} \quad \frac{b - x_1}{b - a} \tag{5.01}$$

while ensuring that $x_1 < x_2$. The best that can be done is to make the two points almost coincident which would then make $l_2 = l_0/2$. Effectively, one evaluation will happen at the midpoint of the interval, and one a small distance to one side. When $N = 3$, two function evaluations must result in an interval where the function is known at the midpoint, and so the first two calculations must correspond to the points which divide the original interval into three equal parts. So $l_3 = l_0/3$.

For dynamic programming, we need to have suitable expressions for the state and stage, together with a recurrence relationship. After we have completed a series of function evaluations, we will have a state which can be summarised by two variables, the length of the interval and the position of the known function value in the interval. The stage is described by the number of times that the function will be evaluated. The decision to be made is the location of the next x−value where $g(x)$ will be calculated and as a consequence of this, there will be a new length of interval and known function value in the interval. So, we define $f_n^*(l,p)$ to be the length of the shortest interval possible given n function calculations with an interval of length l and $g(x)$ known at position p in the interval $(0 < p < l)$. Then, we must choose a new position q at which the function will be calculated. This will provide data for reducing the interval, and as a consequence, the interval at the next stage will be one of the following lengths:

$$p \text{ or } (l-q) \qquad (q<p) \tag{5.02}$$

$$q \text{ or } (l-p) \qquad (p<q) \tag{5.03}$$

The final condition, $f_0^*(l,p) = l$ is a natural consequence of the definition. In consequence, therefore, the recurrence relationship can be written as:

$$f_n^*(l,p) = \min \left\{ \begin{array}{l} \min\limits_{0 \le q < p} \max\big(f_{n-1}^*(p,q), f_{n-1}^*(l-q, l-p)\big) \\ \min\limits_{p \le q < l} \max\big(f_{n-1}^*(q,p), f_{n-1}^*(l-p, l-q)\big) \end{array} \right\} \qquad (n>0) \tag{5.04}$$

This relationship looks somewhat horrendous; this is partly because we have had to consider every possible outcome and then express them in the right hand side of the equation. The solution is not obvious, but can be written

$$f_n^*(l,p) = \min \left\{ \begin{array}{l} \max\Big(\dfrac{p}{F_{n+1}}, \dfrac{l-p}{F_n}\Big) \\ \max\Big(\dfrac{l-p}{F_{n+1}}, \dfrac{p}{F_n}\Big) \end{array} \right\} \tag{5.05}$$

where F_n is the nth Fibonacci number defined by the recurrence

$$\begin{array}{c} F_n = F_{n-1} + F_{n-2} \\ F_0 = 0 \qquad F_1 = 1 \end{array} \tag{5.06}$$

which gives $\{F_0, F_1, F_2, F_3, F_4, F_5, F_6 \ldots\} = \{0, 1, 1, 2, 3, 5, 8, \ldots\}$ (The notation for the Fibonacci numbers is not standardised. In this section we are using that used by Graham *et al* (1988) and in the series of books by Knuth (1968, 1969, 1973).) It is a straightforward process to verify that this solution does satisfy the dynamic programming relationship. Of more interest in practice are the rules which determine where the function should be calculated in order to optimally search the initial interval of length l_0. This is the only time that two function calculations will be needed, and an application of the result above will reveal that these two should be made at positions

$$x_1 = \frac{F_{N-1}}{F_{N+1}} l_0, \qquad x_2 = \frac{F_N}{F_{N+1}} l_0 \tag{5.07}$$

in the initial interval.

So, if this initial interval is $(2.3, 8.8)$ (a length of 6.5 units), and one is allowed 6 function evaluations, the first two should be at $2.3 + \frac{5}{13}6.5, 2.3 + \frac{8}{13}6.5$, that is $4.8, 6.3$ (using $F_7 = 13$). The final interval will be of length 0.5 units. Thus, for example, the function

$$g(x) = |\cos(2x) + \sin(5x) + 7.6x - 26.6| \tag{5.08}$$

(where x is in radians) has its minimum in the interval $(2.3, 8.8)$. The first function evaluations give

$$g(4.8) = 7.9897$$
$$g(6.3) = 22.3634$$

so that the next iteration is concerned with locating the point in the interval $(2.3, 6.3)$ with $g(4.8)$ known; therefore we calculate

$$g(3.8) = 2.6811$$

and the next stage is to locate a point in $(2.3, 4.8)$ with $g(3.8)$ known, and thus we find:

$$g(3.3) = 1.2816$$

Proceeding along these lines one gets: in $(2.3, 3.8)$

$$g(2.8) = 3.5538$$

in $(2.8, 3.8)$

$$g(3.3 - 0.0001) = 1.2819$$

yielding the final interval in which the minimum is known to be located as: $(3.2999, 3.8)$ after 6 function evaluations.

This search is frequently referred to as the "Fibonacci search method" and is the most efficient method of reducing an interval. Its principal drawback is the need to specify the number of times that the function will be calculated before the search begins. An approximation to the method is in more general use, the so-called "Golden Section" method, which uses the limiting value of the ratio $\frac{F_{N-1}}{F_N} = \frac{\sqrt{5}-1}{2} = 0.6180339887$ to determine the placing of the points where the function is calculated. (If this method is applied to the function quoted earlier, with the same starting interval, then after 6 function evaluations, the interval will be reduced to (3.248, 3.834), about 17% longer than with the Fibonacci method.)

Needless to say, the same search procedure can be applied to problems of maximisation. The point where the function $g(x)$ is least is also the point where the function $-g(x)$ is greatest. The procedures are discussed fully by Vajda (1989).

5.3 Transportation problems

A commonly occurring problem in the distribution of goods from manufacturer to customer can be described by the general title of "Transportation problem". The manufacturer wants to distribute goods at minimal cost, which means that different sources of goods need to be associated with the different demand points for the goods in an optimal fashion – where optimal means least cost. It is quite common to suppose that the cost of transporting material between the "source" and the "sink" is proportional to the amount of material being moved, and so the problem of choosing the amount to be carried along the various routes is a linear program.

Dynamic programming offers a different approach to the solution of such problems, and can extend the scope of them from those with linear costs of transport to more general functions. However, as is often the case in dynamic programming, the approach is only suitable for fairly small problems. We shall suppose that there are M depots (which may be factories), which have amounts a_i available ($i = 1, \ldots, M$) and N shops, which have demands b_j for the goods ($j = 1, \ldots, N$) We assume that the total demand is equal to the total amount available (i.e. $\sum_{i=1}^{M} a_i = \sum_{j=1}^{N} b_j$). (Any imbalance can be corrected by introducing a "fictitious" depot or shop.) The cost of transporting x_{ij} units from the ith depot to the jth shop is some known function of the amount transported; for the purpose of this discussion, we shall assume that it is a linear function, $c_{ij}x_{ij}$. Then the transportation problem is the linear programme:

$$\begin{aligned}
\text{Minimise:} \quad & \sum_i \sum_j c_{ij} x_{ij} \\
\text{with:} \quad & \sum_{i=1}^{M} x_{ij} = b_j \qquad \forall j = 1, \ldots, N \\
& \sum_{j=1}^{N} x_{ij} = a_i \qquad \forall i = 1, \ldots, M \\
& x_{ij} \geq 0 \qquad \forall i = 1, \ldots, M \quad j = 1, \ldots, N \quad (5.09)
\end{aligned}$$

The naming of one group of locations as depots and the other as shops does not really affect the form of the problem, since we can exchange the two concepts and have essentially the same problem. The dynamic programming formulation is best suited to situations where the smaller of M and N is only 2 or 3, since beyond this the "curse of dimensionality" becomes important. Our examination of the formulation will concentrate on $M = 2$, although the format of larger problems can be deduced from this one.

Suppose that there are 2 depots, with amounts a_1, a_2 available, and N shops requiring amounts $b_1, b_2, \ldots, b_N$. We suppose that the problem is balanced with total demand equal to the total supply. For each of the $2N$ transport routes we have a cost coefficient c_{ij} for the cost per unit of the goods being transported from the ith depot to the jth shop. The dynamic programme assumes that we make decisions about the quantity of goods that are shipped in sequence, so we shall start with the Nth shop and work backwards to the first one. Let us define a function $f_n^*(x_{1n}, x_{2n})$ to be the cost of an optimal policy for transporting the goods from the two depots given quantities (x_{1n}, x_{2n}) at them when there are the n shops $j = 1, \ldots, n$ to be supplied; then moving y_1 units from depot 1 and y_2 from depot 2 to supply the nth shop will leave quantities $(x_{1n} - y_1, x_{2n} - y_2)$ to supply the remaining $n - 1$ shops. Hence we have a transformation rule and an obvious recurrence relation. The cost of an optimal policy when there are n shops to be supplied will be given by the minimum, taken over all possible amounts, of the sum of the cost of supplying shop number n with goods (the single stage cost, which is linear in the decision variable for the stage) and then supplying the remaining $n - 1$ shops with what remains in an optimal manner. Mathematically, this is written:

$$\begin{aligned} &f_n^*(x_{1n}, x_{2n}) \\ &= \min_{y \in S_n} \Big(c_{1n}y + c_{2n}(b_n - y) + f_{n-1}^*(x_{1n} - y, x_{2n} - (b_n - y)) \Big) \end{aligned} \tag{5.10}$$

where the feasible values of y, those in S_n, are those which ensure that neither the amounts transported nor the amounts left at the depots are negative. So

$$S_n = \{y | y \geq \max(0, b_n - x_{2n}), y \leq \min(x_{1n}, b_n)\} \tag{5.11}$$

Furthermore, the only states which can occur will be those which do not violate the total amounts available for supply, so that $x_{1n} \leq a_1, x_{2n} \leq a_2$

The problem will be simplified because of the balance of supply and demand. When there are no shops to be supplied, then there will be no goods available in either depot; when there is one shop to be supplied, the total quantity available in the depots ($= x_{11} + x_{21}$) will be equal to the demand at this shop, and so specifying either quantity will immediately give the other one. A similar argument extends to all possible (state, stage) pairs of the problem, and as a result, the states can be written as being one-dimensional, not two. In general problems, the state will have P dimensions; $P = \min(M - 1, N - 1)$

This is a valuable simplification of the problem; another, also of great use, follows from the linearity of the single stage costs. As a consequence, the objective function will consist of several linear functions and be piece-wise linear;

hence one will only need to know where the breaks are in this piece-wise linear function. The benefits of this will appear in the example of the application of dynamic programming below.

5.4 Example of the transportation problem

The table (Table 5.1) gives an example of a simple transportation problem. The entries in the table show the costs of moving goods from two depots to each of five shops, and the margins show the amounts of the goods required by the shops and available at the depots.

Table 5.1 Transportation in Wessex

	Shop					
Depot	Ringwood	Salisbury	Trowbridge	Wells	Yeovil	Supply
Blandford	35	43	61	68	43	20
Chippenham	91	60	23	57	81	16
Demand	5	6	7	8	10	Total=36

The final state of the assignment of goods to depots, will have an objective function value 0;

$$f_0^*(0,0) = 0$$

With the recurrence relationship and the simplifying factors already cited, the only states that can occur when goods have to be assigned to the shop at Ringwood will be $(x_{11}, x_{21} = 5 - x_{11})$ for $0 \leq x_{11} \leq 5$ and quite clearly, there will only be one way of arranging delivery. So

$$f_1^*(x_{11}, 5 - x_{11}) = 35x_{11} + 91(5 - x_{11}) = 455 - 56x_{11} \tag{5.12}$$

Working backwards and using the recurrence relation, we find that:

$$\begin{aligned} f_2^*(x_{12}, 11 - x_{12}) &= \min_y \left(43y + 60(6-y) + f_1^*(x_{12} - y, (11 - x_{12}) - (6 - y))\right) \\ &= \min_y \left(815 + 39y - 56x_{12}\right) \end{aligned}$$

where

$$\max(0, x_{12} - 5) \leq y \leq \min(x_{12}, 6) \tag{5.13}$$

To minimise the expression $(815 + 39y - 56x_{12})$, y must be as small as possible. So long as $0 \leq x_{12} \leq 5$, the minimum of this will occur when $y = 0$ and then we will have

$$f_2^*(x_{12}, 11 - x_{12}) = 815 - 56x_{12} \tag{5.14}$$

which is a linear function of the variable x_{12}. However, when $5 \leq x_{12} \leq 11$, the minimum of the expression will occur at $y = x_{12} - 5$ so that

$$f_2^*(x_{12}, 11 - x_{12}) = 815 + 39(x_{12} - 5) - 56x_{12} = 620 - 17x_{12} \qquad (5.15)$$

which is also a linear function of x_{12} but with a different slope. The two linear functions meet at $(5, 535)$. Hence, the value of an optimal policy from stage 2 onwards is the piece-wise linear function illustrated in figure 5.2.

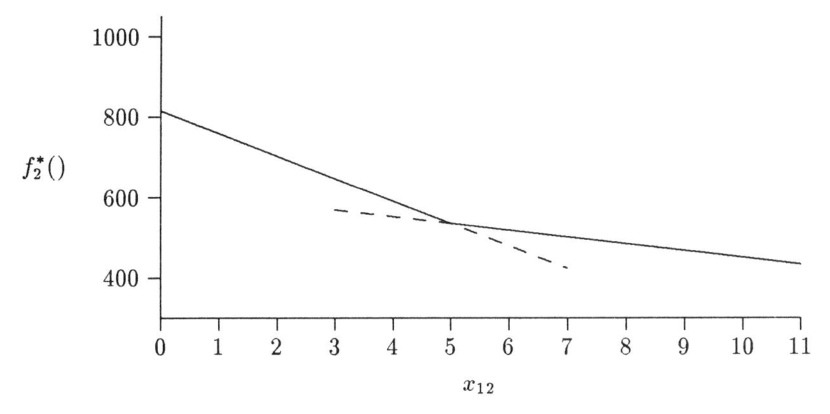

Figure 5.2: The behaviour of $f_2^*(x_{12}, 11 - x_{12})$ as a function of x_{12}

Using the recurrence relationship once more permits us to write down an expression for $f_3^*(x_{13}, 18 - x_{13})$ as follows:

$$f_3^*(x_{13}, 18 - x_{13}) = \min_{\max(0, x_{13}-11) \leq y \leq \min(x_{13}, 7)} (61y + 23(7 - y) + f_2^*(x_{13} - y, 18 - x_{13} - (7 - y))) \qquad (5.16)$$

Replacing $f_2^*()$ by its functional form yields:

$$\begin{aligned} f_3^*(x_{13}, 18 - x_{13}) &= \min_{\max(0, x_{13}-11) \leq y \leq \min(x_{13}, 7)} \left(\begin{cases} 976 + 94y - 56x_{13} & \text{for } 0 \leq x_{13} - y \leq 5 \\ 781 + 55y - 17x_{13} & \text{for } 5 \leq x_{13} - y \leq 11 \end{cases} \right) \\ &= \begin{cases} 976 - 56x_{13} & 0 \leq x_{13} \leq 5, \quad y = 0 \\ 781 - 17x_{13} & 5 \leq x_{13} \leq 11, \quad y = 0 \\ 176 + 38x_{13} & 11 \leq x_{13} \leq 18, \quad y = x_{13} - 11 \end{cases} \end{aligned} \qquad (5.17)$$

which is piece-wise linear and is pictured in Figure 5.3. By the time that stage 3 has been reached, the constraints on the amount of material available at the

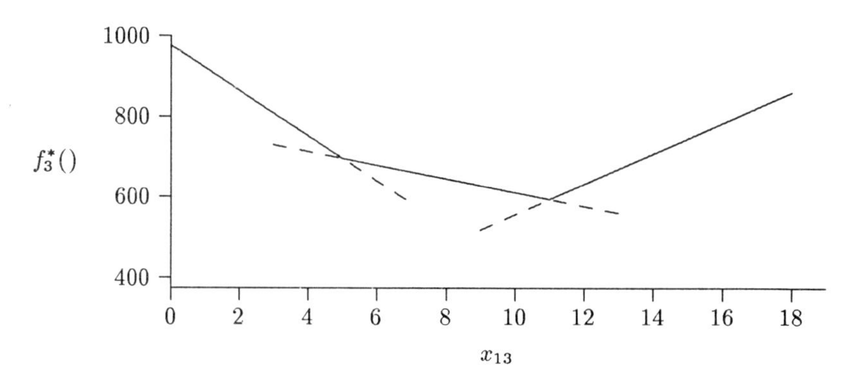

Figure 5.3: The behaviour of $f_3^*(x_{13}, 18 - x_{13})$ as a function of x_{13}

two depots have become important, and we know that the only states that might occur are those with $x_{13} \geq 2$

Working backwards to the fourth stage provides yet another instance of the recurrence relation being used. After the appropriate analysis, the form of $f_4^*(x_{14}, 26 - x_{14})$ will be the four lines:

$$f_4^*(x_{14}, 26 - x_{14}) = \begin{cases} 1432 - 56x_{14} & 0 \leq x_{14} \leq 5, \quad y = 0 \\ 1237 - 17x_{14} & 5 \leq x_{14} \leq 11, \quad y = 0 \\ 929 + 11x_{14} & 11 \leq x_{14} \leq 19, \quad y = x_{14} - 11 \\ 416 + 38x_{14} & 19 \leq x_{14} \leq 26, \quad y = 8 \end{cases} \tag{5.18}$$

ignoring the restrictions on x_{14} from the amounts available at the depots (which would make $10 \leq x_{14} \leq 20$).

Finally, the fifth stage yields the general form:

$$f_5^*(x_{15}, 36 - x_{15}) = \begin{cases} 2242 - 56x_{15} & 0 \leq x_{15} \leq 5, \quad y = 0 \\ 2152 - 38x_{15} & 5 \leq x_{15} \leq 15, \quad y = x_{15} - 5 \\ 1837 - 17x_{15} & 15 \leq x_{15} \leq 21, \quad y = 10 \\ 1249 + 11x_{15} & 21 \leq x_{15} \leq 29, \quad y = 10 \\ 466 + 38x_{15} & 21 \leq x_{15} \leq 36, \quad y = 10 \end{cases} \tag{5.19}$$

In the specific case of concern, $x_{15} = 20$; the optimal allocation at stage 5 will lead through the states (20,16), (10,16), (10,8), (10,1), (5,0), (0,0) to a total transport cost of 1497 units. Interestingly, the dynamic programming formulation provides some information which is not normally available from the solution of a transportation problem. The coefficient of x_{15} in the final expression changes sign at $x_{15} = 21$ If the amount available at the depots were to be varied, the smallest total cost would occur at this point, assuming that all the other terms were to remain unaltered. So, if there were to be scope for sharing the amount of material available in a different way then this would be an optimal allocation. The reason for this is apparent from looking at table 5.1; the supply at the Blandford depot should be allocated to those shops which are closer to it than to the Chippenham depot, which would be Ringwood, Salisbury and Yeovil – a total demand of 21 units.

5.5 Constrained search in many dimensions

Any optimisation problem in n dimensions can be expressed as:

$$\text{maximise} \quad f(x_1, \ldots, x_n) \tag{5.20}$$

subject to

$$(x_1, \ldots, x_n) \in S \subseteq \mathbf{R}^n \tag{5.21}$$

where the set S is defined by a series of constraints or as a set of discrete values of the components of the vector $(x_1, \ldots, x_n)$. There need be no restriction on S; such a problem is unconstrained. In a nonlinear optimisation problem, the objective function $f(\mathbf{x})$ may be nonlinear, or some of the constraints used in the definition of S may be nonlinear, or both.

Dynamic programming is particularly useful when:

(a) the constraints yield a convex region which is contained in a box-set of feasible values bounded above and below;

and

(b) the function can be separated into parts, each of which depends on one variable only, which are added together.

By (a) we mean that each of the n variables, $(x_1, \ldots, x_n)$ has a maximum and a minimum which can be determined without knowing the values of the other $n - 1$ variables. When we do know their values (or the values of some of them), then the bounds may become tighter, but remain in the form of simple limits. (This latter is a result of the convexity of the feasible region). Usually, the lower bounds are assumed to be zero, so that one can talk about the quantity of resource available to the user ..., assuming that this description makes some kind of sense. A convex region is, informally, a region in $\mathbf{R}^n$ such that the whole of the line between any two points $\mathbf{x}$ and $\mathbf{y}$ in the region is also in the region. Textbooks on linear programming will provide a more rigorous definition.

The second condition (b), is a parallel to the problems that were seen earlier where the objective function was calculated by adding together separate parts, each consequent on one decision.

Thus, the region that we are considering is one that can be considered to be inside the n-dimensional equivalent of a box. Each variable x_i has limits l_i and u_i which are known in advance, and $l_i \leq x_i \leq u_i$. Inside this box, the region S is defined by some constraints. This is illustrated in two dimensions in Figure

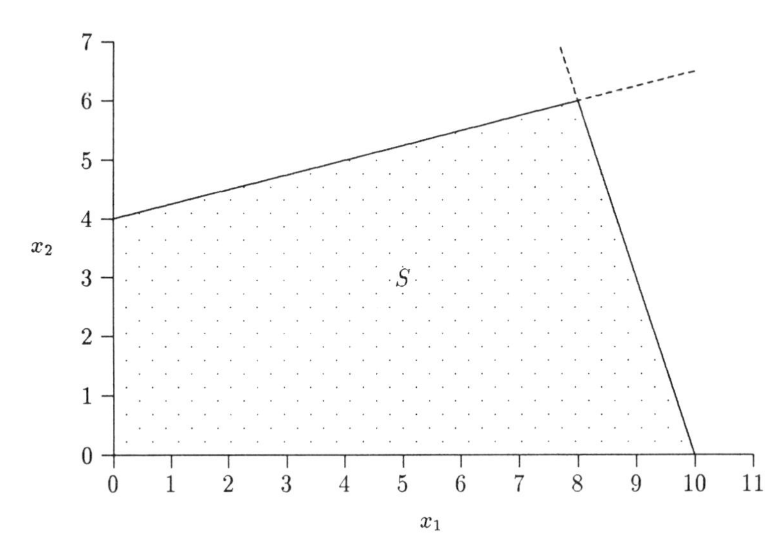

Figure 5.4: The region S

5.4. Here, the convex set S is defined by the four constraints

$$\begin{aligned} x_1 &\geq 0 \\ x_2 &\geq 0 \\ -x_1 + 4x_2 - 16 &\leq 0 \\ 3x_1 + x_2 - 30 &\leq 0 \end{aligned} \tag{5.22}$$

Whatever the value of x_1, x_2 is bounded by the values 0 and 6; whatever the value of x_2, x_1 is bounded by the values 0 and 10. But, once a value for either x_1 or x_2 has been found, then the other one is limited to a (usually) smaller range. Here, if $x_1 = 5$, we deduce that $0 \leq x_2 \leq 21/4$ and if $x_2 = 5$, then $4 \leq x_1 \leq 25/3$.

To apply dynamic programming to a general problem like this, we take each variable in turn, starting with x_n. (It doesn't make any difference to the answer which variable is defined as which; simplicity suggests that one should choose the variable to be called "x_n" that will make for the minimum amount of work ... but that is something which can't be defined readily.) We assume that when we come to optimise with respect to x_n, there is a maximum β_n and a minimum α_n units of resource available as a result of decisions made about the variables $x_1, \ldots, x_{n-1}$. In other words, we have to maximise $f(\mathbf{x})$ subject to $\alpha_n \leq x_n \leq \beta_n$ as a function of x_n alone, keeping the other variables fixed. The answer to this subproblem will appear as a function of s. Once that has been done, we can optimise with respect to x_{n-1} knowing that there is a rule for determining the best that can be achieved for each quantity of resource that is left unused at the end of that stage.

5.6 Example of optimisation

An example will illustrate what is involved. Suppose we have to maximise the function

$$f(x_1, x_2) = x_1^3 - 11x_1^2 + 40x_1 + x_2^3 - 8x_2^2 + 21x_2 \tag{5.23}$$

subject to

$$x_1 + x_2 \leq 6 \qquad x_1, x_2 \geq 0 \tag{5.24}$$

We assume that we have dealt with x_1 and are now concerned to maximise the function with respect to x_2 with an amount of resource which is bounded above by s and below by 0. For fixed x_1, the objective to be maximised is $x_2^3 - 8x_2^2 + 21x_2$ subject to $0 \leq x_2 \leq s \leq 6$ which yields a function value which depends on x_2 and s, say $f_2(s, x_2)$. We are particularly interested in the maximum value of this for a fixed s, say $f_2^*(s)$

By calculus we find that this function has turning values at $x_2 = 7/3$ and $x_2 = 3$; the former of these corresponds to a maximum and the latter to a minimum. (If we had a function that could not be handled by classical methods of optimisation such as the calculus, we would have to use a one-dimensional search method such as the Fibonacci technique described earlier to achieve an acceptable accuracy for the location of the minimum of this one dimensional function.) Then it follows that if $s \leq 7/3$, then the best value of x_2 should be $x_2 = s$. If s is larger, then there will be a set of values of s which run from $s = 7/3$ to $s = t$ for which the best value of x_2 will be $x_2 = 7/3$; t is the solution of the equation

$$(7/3)^3 - 8(7/3)^2 + 21(7/3) = t^3 - 8t^2 + 21t \quad (t > 7/3) \tag{5.25}$$

It is a simple matter to show that $t = 10/3$. Then, if $s \geq 10/3$, the best value of x_2 should again be $x_2 = s$.
Putting all this information together, we see that

$$\begin{aligned} f_2^*(s) &= s^3 - 8s^2 + 21s \quad (0 \leq s \leq 7/3) \\ &= 490/27 \quad (7/3 \leq s \leq 10/3) \\ &= s^3 - 8s^2 + 21s \quad (s \geq 10/3) \end{aligned} \tag{5.26}$$

Working backwards to the problem of optimising the function

$$f_1(s, x_1) = x_1^3 - 11x_1^2 + 40x_1 + f_2^*(s - x_1) \tag{5.27}$$

which must be maximised as a function of x_1 for given amount of resource s. The example has $s = 6$. Calculus fails to answer all the questions here; f_1 is not differentiable at all points, so the problem must be broken into smaller pieces;

$$0 \leq s - x_1 \leq 7/3 \tag{5.28}$$

$$7/3 \leq s - x_1 \leq 10/3 \tag{5.29}$$

$$10/3 \leq s - x_1 \leq 6 \tag{5.30}$$

In the first and third intervals, the function to be maximised can be written as

$$f_1(s, x_1) = x_1^3 - 11x_1^2 + 40x_1 + (s - x_1)^3 - 8(s - x_1)^2 + 21(s - x_1) \quad (5.31)$$

which has a turning value at

$$x_1 = \frac{19 + 16s - 3s^2}{38 - 6s} \quad (5.32)$$

We are concerned about the behaviour when $s = 6$; then the turning value is at $x_1 = 5.5$ and $x_2 = 0.5$; however the dynamic programming formulation allows us to consider the more general problem of the optimum for any value of resource s. The function value at this turning point is 62.25.

In the second interval, the function to be maximised is simpler:

$$f(s, x_1) = x_1^3 - 11x_1^2 + 40x_1 + 490/27 \quad (5.33)$$

whose turning points are at $x_1 = 10/3$ (a local maximum in the interval when $s = 6$) and $x_1 = 4$ (a local minimum which lies outside the interval when $s = 6$). The function value at $x_1 = 10/3$, $x_2 = 7/3$ is 66.296296; this then is the solution to the optimisation problem.

The method can be extended to problems with more than two variables and to more complex functions. In general applications are quite limited and the growing technology for nonlinear optimisation limits the value of the dynamic programming formulation. There is a parallel between this method and the problems of resource allocation from Chapter 3, where the discrete nature of the decision variables meant that the nonlinear objective function was represented by a table of values and not in a functional form.

5.7 Exercises

(1) Use the Fibonacci search method to locate more accurately that minimum of

$$g(x) = \sin(x) + \exp(x) - x$$

which lies in $(-1.0, -0.79)$ using 8 function evaluations.

(2) Verify (do not prove) the assertion that the points determined by the Fibonacci numbers satisfy the recurrence relationship for one-dimensional optimisation.

(3) Solve the transportation problem shown in Table 5.2 using dynamic programming.

(4) Solve the transportation problem of Table 5.3 using dynamic programming.

(5) Formulate the transportation problem of Table 5.4 as a dynamic programme, but do not solve it.

Table 5.2: Distances for exercise 3.

	Shop				
Depot	Aberdare	Brecon	Cardiff	Denbigh	Available
Newport	50	65	43	100	5
Porthcawl	48	84	23	78	7
Required	2	4	4	2	Total=12

Table 5.3: Distances for exercise 4

Depot ↓ Shop →	St Andrew's	Tarbet	Available
Crieff	67	70	5
Dunoon	166	55	8
Edinburgh	79	102	7
Forfar	38	142	6
Required	14	12	Total=26

Table 5.4: Distances for exercise 5

Depot ↓ Shop →	Chertsey	Datchet	Esher	Available
Hampton	14	11	9	5
Isleworth	7	7	13	7
Kew	10	10	13	11
Required	6	8	9	Total=23

(6) Use the results of the example of two-dimensional optimisation in the chapter to solve the problem: maximise the function

$$f(x_1, x_2) = x_1^3 - 11x_1^2 + 40x_1 + x_2^3 - 8x_2^2 + 21x_2$$

subject to

$$x_1 + x_2 \leq 5.4$$
$$x_1, x_2 \geq 0$$

(7) Using dynamic programming, maximise the function

$$f(x_1, x_2) = x_1^3 - 11x_1^2 + 40x_1 + x_2^3 - 8x_2^2 + 21x_2$$

subject to

$$x_1 + x_2 \leq 6.6$$
$$x_1, x_2 \geq 0$$

(8) Gongola Foods makes two vegetable sauces, which have recently been packaged in a new format. The company's advertising department has been given a budget of £1 million to publicise these products, and is about to allocate this sum between different media. The evidence from market research is that the response will depend on the amount spent according to the following four functions:

1) Television cover for courgette and tomato sauce given expenditure of x_1 : $-x_1^2 + 7x_1$
2) Magazine advertising for courgette and tomato sauce given expenditure of x_2: $-x_2^2 + 3x_2 + 2$
3) Television cover for bean and mushroom sauce given expenditure of x_3 : $-2x_3^2 + 6x_3$
4) Magazine advertising for bean and mushroom sauce given expenditure of x_4: $-x_4^2 + 10x_2 + 4$

(expenditure is measured in units of £100 000; the effects of not advertising have been included in the functions for magazine advertising.)

Formulate the decision problem as a dynamic programme, and find the solution which maximises the response. How sensitive is the solution to the quadratic coefficients?

6

Randomness in dynamic programming

6.1 Introduction

So far in this book, all the models examined have been deterministic ones. These have been chosen because they are simpler and less complex to manipulate. However, they are also usually less than adequate for modelling the real world, where chance and randomness are important features. To be better modellers of practical situations, it is necessary to take into account the stochastic nature of the real world in some way or other. The Operational Research Society in the United Kingdom used a definition of O.R. for several years which read, in part:

> "... The distinctive approach is to develop a scientific model of the system, incorporating measurements of factors such as chance and risk, with which to predict and compare the outcomes of alternative decisions, strategies or controls. ..."

"Chance" and "risk" are associated with randomness.

The deficiencies (in some cases) of our models have been because the functions that we have considered have been known precisely. Every transformation between states has been deterministic. One mathematical consequence of this has been that the way that stages have been ordered has been potentially arbitrary; in many cases it would be possible to solve the dynamic programming problem by working from start to finish rather than from finish to start. If one did this, using the method known as "forward recurrence", the decisions to be made in a state would be how to approach the state, and not how to leave it.

In knapsack problems, the ordering of the items to be included did not affect the solution, and it would be possible to choose an order to attempt to minimise the work associated with solving the recurrence relationships. In the second part of this book, we shall consider problems where chance and risk are present and in which the stages must be ordered in a particular way. The outcome of a decision made in a particular state at a particular stage involves some unpredictability; there will be a probability distribution describing the possible states which will be encountered at the next stage.

A very simple example can be used to illustrate this. Suppose that you are gambling against a generous opponent who describes a game with a fair die as follows: you can roll the die up to three times; when you have seen the value on the die after the first roll, X_1, you decide whether to roll it again or not. If you decide to stop, then you will be paid £X_1. If you roll the die a second time, the result will be X_2, and you have to decide whether or not to continue, with a reward of £X_2 if you stop. If you roll the die for a third time, scoring X_3, you will receive £X_3. You want to play in the best way possible, which is interpreted as an objective of maximising your expected income from the game. How should you proceed? It will be clear that it is important to follow the sequence of decisions in their logical order; the decision when the die has been rolled for the second time implies that the decision when it was first rolled has already been taken.

After the first roll (and the second if you continue), you will be faced with a decision: stop or continue. These two decision points correspond to stages in a sequential process, whose states are defined in terms of the score (X_i, which takes values 1, ..., 6 for each of $i = 1, 2, 3$) in front of you when you make a decision. If you continue, the state which you will reach at the next stage depends on the roll of the die, which is random. All six states have the same probability of occurring.

Later in this chapter we shall return to consider the example and seek an optimum policy for our decision.

6.2 Some fundamental ideas

Our basic assumptions will not be changed significantly, beyond the introduction of randomness. Our study will be of a model in which there is a succession of choices to be made, one after the other. Although decisions in practical situations may be made at any time, we shall restrict our consideration to cases where the set of times when decisions may be made is limited. These times will form the stages, indexed in the same way as in other cases, of a dynamic programming formulation. The state at stage j will be a description of the system, which will be represented as a random variable (or possible a vector of variables) X_j. At stage j, we have a range of possible decisions, identified as 1,2,...,D_j. Our objective is to choose the best of these, d_j^*. The optimal decision will depend on the state X_j and the stage. Once we have made a decision, then the system moves onward to a new state X_{j+1} at stage $j + 1$. Therefore, as time progresses, the system experiences an ordered series of states and decisions:

$$X_1, d_1^*, X_2, d_2^*, X_3, d_3^*, \ldots, X_j, d_j^*, \ldots$$

The whole series may only be observed after the event. When we are actually making the decisions, only some of this information is known. The knowledge at stage j will be in two parts; the certain "past", what has happened up to and including the present state; and the uncertain "future" states. Because of the stochastic nature of the problem, we only have ideas of what may happen in subsequent stages. These ideas may be in the form of an indication of the probability distribution of the states which might be reached by making decisions from the present one. In general there are various levels of knowledge which might go with this probability distribution; in some cases, it might be completely unknown; in others, we might have some idea about the distribution, but not about the parameters of it; the best that we can expect is to know the distribution and its parameters exactly.† Our discussion will concentrate on this last level of knowledge.

With a probability (density) function, it is possible to express the next state X_{j+1} as a function of the series of states and decisions which have come before (the "past"), the present state and the decision that is taken. So the probability function (for discrete states) will be in the form

$$\text{prob } (X_{j+1} = x_{j+1} \text{ given decision } d_j) = \phi(x_{j+1} | X_1, d_1^*, X_2, d_2^*, \ldots, X_j, d_j) \qquad (6.01)$$

(and there will be a corresponding density function for continuous distributions).

When we attack such problems with dynamic programming we shall restrict ourselves to those functions which are independent of all the past states. Then our probability function will be much simpler:

$$\text{prob } (X_{j+1} = x_{j+1} \text{ given decision } d_j) = \theta(x_{j+1} | X_j, d_j) \qquad (6.02)$$

Such situations are Markovian; that is, they have the Markov property that only the present state of the system affects its future behaviour. The states of the system at successive stages represent a Markov process whose transition

† As examples of these three situations, the first might be illustrated by a state "The day when this newly planted tree first produces fruit"; the second by the score in a football match tomorrow; the third by a system such as our initial example, where the probabilities of all states are known exactly.

probabilities are determined by the decisions that are taken. One of the alternative statements of Bellman's principle assumed that this would be the case for an optimal policy.

In this chapter the Markov processes to be considered will be ones which have a finite horizon; there is no possibility of them extending into the future beyond a particular stage, which is known to occur after a finite number of decisions. The example of the die rolling game falls into this category, as the game stops after at most three rolls of the die. (One could envisage a game with the same rules, in which one went on indefinitely; the best thing to do then is to stop when a six has come up.) We shall suppose that the system being modelled has a maximum of N stages, identified as $1, 2, \ldots, N-1, N$.

6.3 Markov processes with rewards

Fundamental to the dynamic programming process is the choice of a recurrence relationship. In the discussion of deterministic dynamic programming, several different formulations have emerged. When there is randomness, then there are yet more potential descriptions of the functional link between the values of the objective function at successive stages. In this section we shall start to look at a few of these.

Very often, the intention of a decision-maker is to maximise a reward, which is analagous to minimising a cost. Then it is sensible to have the objective in dynamic programming as the reward from the states, decisions and transitions; in the case of random variation, the objective will reflect the stochastic transitions by using the expected value of the reward.

Suppose that at stage j the system is in stage X_j; a decision is made, d_j, and as a result of this and the inherent randomness, the state at the next stage is X_{j+1}, a state which is determined by some probability function (a density function or a discrete probability function) $\phi(X_{j+1}|X_j, d_j)$. The transition and the decision lead to a reward $K(X_j, d_j, X_{j+1})$. From this, it is possible to find the expected reward from making decision d_j, $EK(X_j, d_j)$, in state X_j by integration or summation as appropriate. The reward from making a policy decision d_j followed by optimal decisions thereafter will be:

$$EK(X_j, d_j) + \text{expected return, with optimal policy from stage } j+1 \text{ on } |(X_j, d_j)$$

Let us denote this by $f_j(X_j, d_j)$ Then our mathematical expression of the dynamic programming recurrence relationship becomes:

$$f_j(X_j, d_j) = EK(X_j, d_j) + \sum_{\text{all } x_{j+1}} \left(\phi(x_{j+1}|X_j, d_j) f^*_{j+1}(x_{j+1})\right) \qquad (6.03)$$

where $f^*_{j+1}(x_{j+1})$ is the return from an optimal policy from state x_{j+1} at stage $j+1$. This return is identified by taking the maximum of the function over all possible decisions, so that the optimal policy in state j is found by solving

$$f_j^*(X_j) = \max_{\text{all decisions } d_j} \left(f_j(X_j, d_j)\right) \tag{6.04}$$

This may seem somewhat involved. The mathematics is, but the concepts are not. What we are trying to establish is the best decision to make in a given state of the system X_j at stage j. We have a choice of decisions, and we identify the one we take as d_j. As a result of this decision, two things happen: first, a set of transition probabilities comes into effect, which will transform the state from its present one to a new one at the next stage; second, a set of rewards becomes relevant, which describe the interconnected costs and benefits of this transition and the decision. Because the transformation is random, the actual return from the stage (the "single-stage return") is a random variable, but we can identify its expected value by simple probability rules. When we reach the next stage, we don't know exactly which state we shall be in, but we know that we shall follow a set of optimal decisions thereafter. Therefore we can calculate the expected return from that stage to the time horizon (or the last stage of all) by applying the same basic probability rules, since we have the probabilities of the possible states that the system could occupy at stage $j+1$.

All that remains is to work backwards from the simplest case, which is what to do in the final stage; generally, finding the values of $f_N^*(X_N)$ for all X_N is a straightforward matter. Once this has been done, then the recurrence relationship can be applied to states $N-1, N-2, N-3, \ldots, 3, 2, 1$; so the problem can be solved.

As an example of this problem, we can look at the die rolling game introduced earlier.

6.4 The die rolling game

In order to analyse the game with the cubical die presented at the start of the chapter, we need to define the standard dynamic programming concepts for such a decision problem. Some were intimated in the description of the problem and the short discussion given there. The problem has three stages and each one has six states. The stages correspond to the first, second and third rolls of the die and the states to the score which is showing on the die. The decision to be made each time is the simple binary (meaning there are only two options) choice between continuing with the game (a decision which incurs no cost and receives no direct reward) and stopping with the reward that corresponds to the present score. At the last stage, this choice exists, but the optimal decision is a trivial one, to stop and take the reward £X_3. Our concept of optimality is that we aim to maximise the expected reward we take away at the end of the game.

So, let us define $f_n^*(X_n)$ to be the expected value of the reward under an optimal policy, taken at the nth roll of the die, when the score showing is X_n.

f_n^* is itself a random variable, because it depends on X_n, and so it too has a probability distribution and an expected value, $E(f_n^*)$. $f_n^*(X_n)$ can take two values: X_n if you stop, $E(f_{n+1}^*)$ if you go on; clearly:

$$f_n^*(X_n) = \max(X_n, E(f_{n+1}^*)) \tag{6.05}$$

with $E(f_4^*)$ defined to be 0.

We are now in a position to answer the problem of the optimal policy. Suppose that we have reached the third roll of the die. The best policy will be determined by solving:

$$f_3^*(X_3) = \max(X_3, 0)$$

which will result in

$$f_3^*(X_3) = X_3$$

Thus f_3^* has six equally likely outcomes, and the expected value of f_3^* will be

$$E(f_3^*) = \frac{1+2+3+4+5+6}{6} = 3.5$$

With this information, we can move back to the second decision, when X_2 is showing (assuming that we get that far). We must solve:

$$\begin{aligned} f_2^*(X_2) &= \max(X_2, E(f_3^*)) \\ &= \max(X_2, 3.5) \end{aligned}$$

This will be solved by letting $f_2^*(X_2) = X_2$ when $X_2 = 4, 5$ or 6 and $f_2^*(X_2) = E(f_3^*)$ otherwise. The best decision that can be made, if the die has been rolled a second time, is to stop if it is showing 4, 5 or 6 and to continue if it shows 1, 2 or 3. The expected value of f_2^* will be

$$E(f_2^*) = \frac{(3.5 \times 3)}{6} + \frac{4}{6} + \frac{5}{6} + \frac{6}{6} = 4.25$$

It is now possible to move back to the first decision, when X_1 is showing on the die. The return under an optimal strategy will be found by solving:

$$\begin{aligned} f_1^*(X_1) &= \max(X_1, E(f_2^*)) \\ &= \max(X_1, 4.25) \end{aligned}$$

The solution is similar to that for the second roll of the die. If the score which is visible is 1, 2, 3 or 4, then the game should continue, because you expect to gain £4.25 by doing that; if you see a 5 or a 6, then you should stop and not continue (even though it is possible that you might do better later on). The expected value of f_1^* is

$$E(f_1^*) = \frac{(4.25 \times 4)}{6} + \frac{5}{6} + \frac{6}{6} = 4.667$$

This represents the amount that you would expect to win, on average, if you played the game in an optimal way; thus, to be fair, £4.667 is the entry fee that should be paid to play the game.

The solution can be used to provide the probability distribution of the amount that will be taken away by a player of the game who follows the optimal rules. There are 86 possible results of the game: 2 where the die is rolled once, 12 where it is rolled twice and 72 where three rolls are needed. The probability of taking away £1, £2 and £3 are all equal to 1/18. For £4 the probability rises to 1/6 since this reward can occur on the second roll of the die under an optimal policy. For the other two rewards, £5 and £6, the probabilities are each equal to 1/3. So, having paid the £4.667 entry fee, on average 2 games from 3 will see a prize greater than this.

Problems such as this are sometimes known as "secretary" problems, when they take their name from a scenario that reflects a somewhat sad sexual stereotype. The setting is of a male office manager interviewing (female) applicants for the post of secretary. The applicants appear one at a time, and the skills of the jth are assumed to be a random variable X_j. The manager must decide whether or not to hire an applicant on the spot; if he turns one down, it is not possible to go back for her later, however much he may regret his decision. What is the best policy for the manager to adopt in order to maximise the expected rating of the skills in the person he employs? Seen in such a light, they are clearly representative of a number of practical decision problems where a sequence of courses of action are presented and there is no scope for reconsidering them later.

Extensions and alterations to this simple example are numerous, and some are provided as exercises at the end of the chapter.

6.5 Inventory control problems

A second situation in which stochastic dynamic programming becomes of interest to the operational research scientist is that of inventory control. Earlier, we examined a model for the case where the future orders were known exactly, and the decision was whether or not to produce goods during a particular period. A similar situation arises in problems where the future demand is random. We consider the problem of a manufacturer who makes decisions at the start of time periods, $t = 1, 2, 3, \ldots, N$, about the number of items to make during the period of time before the next opportunity for a decision. During a time period, $t = n$, there will be demand for D_n items; this will be a random variable, and we shall assume that tere is a probability function $p_{n,k} = \text{probability}\ (D_n = k) =$ gives the distribution of this demand. It may vary with time, to allow for seasonal changes in demand. Costs arise from several sources, as in the case of deterministic demand: there will be the cost of initiating a production run, F_n, the cost of producing X items, $P_n X$ (assuming linearity for simplicity), a cost per item of inventory held from one period to the next, h_n, and (possibly) a cost of running out of stock.

For a retailer, many of these costs will also be relevant. In addition there may well be a selling price per item, leading to a figure for the profit from sales. Alternatively, the objective may be to minimise the total cost of operating the production and inventory system over a specified number of time periods N.

The decisions that are to be made at the start of each time period will be about the number of items to make or buy; after the decision, one will be able to calculate the cost of the operation for the time period, and there will be a random number of items left unused at the end of it. So we can find the distribution of the inventory size at the time when the next decision has to be made. The information which will help the decision-maker is the size of the inventory at the start of the period, and so in dynamic programming terms, we can use this as the state variable, while the period number will be the stage. All that remains to be considered is the form of the recurrence relationship. There are several candidates, depending on the factors which are important for the decision-maker. In keeping with the general format of objective functions, we shall denote each alternative as:

$$f_n^*(I_n) = \text{expected cost of an optimal policy starting at stage } n \text{ with inventory } I_n \tag{6.06}$$

and

$$f_n^*(I_n, x) = \text{expected cost of an optimal policy starting at stage } n \text{ with inventory } I_n, \text{ producing } x \text{ items and following optimal rules thereafter} \tag{6.07}$$

1) If the decision maker wants to minimise costs while always having items in stock, then

$$f_n(I_n, 0) = h_n E(I_n - D_n) + E\big(f_{n-1}^*(I_n - D_n)\big) \tag{6.08}$$

$$f_n(I_n, x) = F_n + P_n x + h_n E(I_n - D_n + x) + E\big(f_{n-1}^*(I_n - D_n + x)\big) \quad (x > 0) \tag{6.09}$$

$$f_n^*(I_n) = \min_{I_n - D_n + x \geq 0} \big(f_n(I_n, x)\big) \tag{6.10}$$

with $I_n \geq 0$.

Equation (6.08) states the obvious, that the cost of a decision to produce nothing in the next time period is zero, but there will be a random number of items to be held over until the next time period, $I_n - D_n$. After that, an optimal policy should be followed, with this as the initial inventory. Equation (6.09) deals with the case where x items are produced, thereby incurring the fixed costs of a production run and the variable costs which depend on the size of the run. In both cases, there is an inventory at the end of the time period which is a random variable and this incurs a holding charge as well as being the opening inventory for the next time interval. The formulations use the expected value of the inventory to calculate the holding costs, and the expected value of an optimal policy from stage $n + 1$. Finally, equation (6.10) shows how the optimal policy from stage n is evaluated as the minimum over all permitted values of x. The condition on x is expressed in the form $I_n - D_n + x \geq 0$, which has to be rearranged to give $x \geq D_n - I_n$ for all possible values of the random variable D_n, so that we must have: $x \geq \max(D_n) - I_n$.

2) If the decision-maker wants to minimise costs, but is willing to be out-of-stock and lose demand, then the recurrence relationships will change to reflect this possibility. Suppose that there is a charge (or penalty) of π_n per item out-of-stock in period n. Then:

$$\begin{aligned} f_n(I_n, 0) &= h_n E(\max(I_n - D_n, 0)) \\ &\quad + \pi_n E(\max(0, D_n - I_n)) \\ &\quad + E\big(f^*_{n-1}(\max(I_n - D_n, 0))\big) \qquad (6.11) \\ f_n(I_n, x) &= F_n + P_n x + h_n E(\max(I_n - D_n + x, 0)) \\ &\quad + \pi_n E(\max(0, D_n - I_n - x)) \\ &\quad + E\big(f^*_{n-1}(\max(I_n - D_n + x, 0))\big) \qquad (x > 0) \qquad (6.12) \\ f^*_n(I_n) &= \min_{x \geq 0}\big(f_n(I_n, x)\big) \qquad (6.13) \end{aligned}$$

with $I_n \geq 0$.

The obvious difference is that we have taken the inventory at the end of the period as either 0 or $I_n - D_n + x$, whichever is the larger, and have brought in terms to describe the penalty cost of the lost sales, if it arises. There is then no restriction on the decision variable x, other than that it should be non-negative.

3) When the objective is to minimise costs, but there is the possibility of backordering items when the inventory level falls to zero, at a charge of b_n per item per time period, the recurrence relationships are similar to those in 2):

$$\begin{aligned} f_n(I_n, 0) &= h_n E(\max(I_n - D_n, 0)) \\ &\quad + b_n E(\max(0, D_n - I_n)) \\ &\quad + E\big(f^*_{n-1}(I_n - D_n)\big) \qquad (6.14) \\ f_n(I_n, x) &= F_n + P_n x + h_n E(\max(I_n - D_n + x, 0)) \\ &\quad + b_n E(\max(0, D_n - I_n - x)) \\ &\quad + E\big(f^*_{n-1}(I_n - D_n + x)\big) \qquad (x > 0) \qquad (6.15) \\ f^*_n(I_n) &= \min_{x \geq 0}\big(f_n(I_n, x)\big) \qquad (6.16) \end{aligned}$$

The important difference is that the possible states of the system now include the possibility of having a negative inventory level, which increases the number of possible states of the system.

In all these examples, it will be necessary to specify what will happen at the end of the planning period. The model should include rules that give the rewards of either disposing of surplus goods, or of meeting the backordered demand from the last time period. This often means extending the model to incorporate an extra stage, the fictitious final stage when imbalances are sorted out.

These three possibilities should be viewed as examples only; there will be further possible cost functions when considering profits in other environments.

6.6 Inventory examples

In the period before a holiday, the manager of a small sweet-shop is planning the policy that he should adopt for stocking a particular brand of chocolates. Demand over the next four weeks is expected to be random, following the pattern shown in Table 6.1:

Table 6.1: Demand for chocolates

Week (n)	1	2	3	4
$\text{Prob}(D_n = 0) = p_{n,0}$	0.18	0.22	0.15	0.10
$\text{Prob}(D_n = 1) = p_{n,1}$	0.31	0.33	0.28	0.23
$\text{Prob}(D_n = 2) = p_{n,2}$	0.26	0.25	0.27	0.27
$\text{Prob}(D_n = 3) = p_{n,3}$	0.15	0.13	0.17	0.20
$\text{Prob}(D_n = 4) = p_{n,4}$	0.06	0.05	0.08	0.12
$\text{Prob}(D_n = 5) = p_{n,5}$	0.04	0.02	0.05	0.08

What policy should the manager adopt if he wants to meet all anticipated demand? Stockholding costs £0.25 per item per week, and the fixed cost of replenishing stock is £5.00; items are available at £2.50 each on sale-or-return, but it will cost a fixed sum of £5.00 to return any items which are left at the end of the $4th$ week.

Let us define $f_n^*(s)$ to be the cost of an optimal policy assuming that at the start of week n the number of items in stock is s, and $f_n(s, x)$ to be the cost of a policy which makes the decision "buy x items in week n and follow an optimal policy thereafter". The recurrence relationship will correspond to those given in equations (6.08)-(6.10) above.

We should start with a function for the fictitious week 5 of the period, to determine what is the best thing to do with the leftover goods. So we define $f_0^*(s)$ to be the cost of an optimal policy when the options are to throw the goods away or to return them to the supplier. This could be determined by common sense – if you have one item left at the end of the period, then it is not worth taking it back; if you have two, then it is as cheap to take them back as not; if you have more than two, then the best thing to do is to take them back. This emerges mathematically from the equations:

$$f_0^*(s) = \min(5.00 - 2.50s, 0) \qquad (s \geq 0) \tag{6.17}$$

so that we have the returns shown in Table 6.2.

Table 6.2: Return from optimal policy for goods that are unsold

s	$f_0^*(s)$ (in £s)
0	0.00
1	0.00
2	0.00
3	-2.50
4	-5.00
5	-7.50
etc	5.00-2.50s

We can go on to $f_1^*(s)$ Thus:

$$f_1^*(s) = \min_{x \geq (5-s)} \Big[5.00 + 2.50x + \sum_{d=0}^{5} \big(p_{4,d}(0.25(s+x-d) + f_0^*(s+x-d))\big)\Big] \qquad (s \geq 0) \tag{6.18}$$

yielding the returns shown in Table 6.3.

In this table and in those that follow, the figures in the second column show the possible values of $f_1(s,x)$ for $x = (5-s), (6-s), \ldots$. To illustrate the calculation, consider $f_1^*(3)$ The recurrence relation, (6.18), takes the form:

$$f_1^*(3) = \min_{x \geq 2} \Big[5.00 + 2.50x + \sum_{d=0}^{5} \big(p_{4,d}(0.25(3+x-d) + f_0^*(3+x-d))\big)\Big]$$

Expanding this for the possible values of x, one has:

$$f_1^*(3) = \min\Big[10.00 + \sum_{d=0}^{5} \big(p_{4,d}(0.25(5-d) + f_0^*(5-d))\big), \qquad (= 8.11)$$

Table 6.3: Returns and orders (week 4)

s	$f_1^*(s)$ in £s	Optimal order
0	min(15.61, 16.36, 16.81, 17.06, 17.31, 17.56) = 15.61	5
1	min(13.11, 13.86, 14.31, 14.56, 14.81, 15.06, 15.31) = 13.11	4
2	min(10.61, 11.36, 11.81, 12.06, 12.31, 12.56, 12.81, 13.06) = 10.61	3
3	min(8.11, 8.86, 9.31, 9.56, 9.81, 10.06, 10.31, 10.56, 10.81) = 8.11	2
4	min(5.61, 6.36, 6.81, 7.06, 7.31, 7.56, 7.81, 8.06, 8.31, 8.56) = 5.61	1
5	min(-1.89, 3.86, 4.31, 4.56, 4.81, 5.06, 5.31, 5.56, 5.81, 6.06, 6.31) = -1.89	0
6	min(-3.64, 1.81, 2.06, 2.31, 2.56, 2.81, 3.06, 3.31, 3.56, 3.81, 4.06) = -3.64	0
7	min(-5.69, -0.44, -0.19, 0.06, 0.31, 0.56, 0.81, 1.06, 1.31, 1.56, 1.81) = -5.69	0
8	min(-7.94, -2.69, -2.44, -2.19, -1.94, -1.69, -1.44, -1.19, -0.94, -0.69, -0.44) = -7.94	0
9	min(-10.19, -4.94, -4.69, -4.44, -4.19, -3.94, -3.69, -3.44, -3.19, -2.94, -2.69) = -10.19	0
10	min(-12.44, -7.19, -6.94, -6.69, -6.44, -6.19, -5.94, -5.69, -5.44, -5.19, -4.94) = -12.44	0

$$= 12.50 + \sum_{d=0}^{5} (p_{4,d}(0.25(6-d) + f_0^*(6-d))), \qquad (= 8.86)$$

$$= 15.00 + \sum_{d=0}^{5} (p_{4,d}(0.25(7-d) + f_0^*(7-d))), \qquad (= 9.31)$$

$$= 17.50 + \sum_{d=0}^{5} (p_{4,d}(0.25(8-d) + f_0^*(8-d))), \qquad (= 9.56)$$

$$= 20.00 + \sum_{d=0}^{5} (p_{4,d}(0.25(9-d) + f_0^*(9-d))), \qquad (= 9.81)$$

$$= 22.50 + \sum_{d=0}^{5} (p_{4,d}(0.25(10-d) + f_0^*(10-d))), \qquad (= 10.06)$$

$$= 25.00 + \sum_{d=0}^{5}(p_{4,d}(0.25(11-d) + f_0^*(11-d))), \qquad (= 10.31)$$

$$= 27.50 + \sum_{d=0}^{5}(p_{4,d}(0.25(12-d) + f_0^*(12-d))), \qquad (= 10.56)$$

$$= 30.00 + \sum_{d=0}^{5}(p_{4,d}(0.25(13-d) + f_0^*(13-d))) \qquad (= 10.81)]$$

One week earlier, when we consider $f_2^*(s)$, it is again clear that when s is less than 5, it is essential to buy some items.

$$f_2^*(s) = \min_{x \geq (5-s)} \Big(5.00 + 2.50x + \sum_{d=0}^{5}(p_{3,d}(0.25(s+x-d) + f_1^*(s+x-d)))\Big) \qquad (s \geq 0)$$

yielding Table 6.4.

The common pattern that emerges from these values is that the best policy is to order sufficient items that the inventory level is exactly 8 items if one has to place an order ($s \leq 4$) and not to order at all if one has sufficient items to meet all possible demands during the coming week.

Working backwards to $f_3^*(s)$, the recurrence relationship yields Table 6.5 in which the common pattern is that one should either order items to bring the inventory level to 9 (for $s \leq 4$) or not at all. A similar pattern appears when we reach the returns for optimal decisions in the first week in Table 6.6.

The policy now is to order to an inventory level of 10 items unless one has sufficient in stock to meet all possible demands in the first week.

The optimal policy which emerges from all the stages of the formulation is that one ought to order enough items to bring the total inventory up to some constant level. Closer examination of the figures in each of the tables reveals that the expected costs of alternative policies differ by a consistent amount from the optimal one. Thus, in the last table, $f_4^*(2) = £31.25$, and this corresponds to an order size of 8 items. The expected cost of ordering one extra item is £31.34, only £0.09 more. This same figure of £0.09 occurs as the difference between the expected cost of ordering 8 items when $s = 3$ and the optimal order size of 7 items, and it appears again in each of the other rows of the table where $s \leq 4$. Similarly, for this range of values of s, the difference between the expected cost of ordering one item too few and ordering the optimal number of items is always £0.29. The tables also show that when $s \geq 5$, the expected cost of a policy has a local minimum at the value one might anticipate, that of ordering up to a given inventory level. So the figures in parentheses in the row for $f_4^*(7)$ are

Table 6.4: Returns and orders (week 3)

s	$f_2^*(s)$ in £s	Optimal order
0	min(25.39, 24.35, 23.53, 23.29, 23.50, 23.81) = 23.29	8
1	min(22.89, 21.85, 21.03, 20.79, 21.00, 21.31, 21.85) = 20.79	7
2	min(20.39, 19.35, 18.53, 18.29, 18.50, 18.81, 19.35, 19.86) = 18.29	6
3	min(17.89, 16.85, 16.03, 15.79, 16.00, 16.31, 16.85, 17.36, 17.86) = 15.79	5
4	min(15.39, 14.35, 13.53, 13.29, 13.50, 13.81, 14.35, 14.86, 15.36, 15.86) = 13.29	4
5	min(7.89, 11.85, 11.03, 10.79, 11.00, 11.31, 11.85, 12.36, 12.86, 13.36, 13.86) = 7.89	0
6	min(4.35, 8.53, 8.29, 8.50, 8.81, 9.35, 9.86, 10.36, 10.86, 11.36, 11.86) = 4.35	0
7	min(1.03, 5.79, 6.00, 6.31, 6.85, 7.36, 7.86, 8.36, 8.86, 9.36, 9.86) = 1.03	0
8	min(-1.71, 3.50, 3.81, 4.35, 4.86, 5.36, 5.86, 6.36, 6.86, 7.36, 7.86) = -1.71	0
9	min(-4.00, 1.31, 1.85, 2.36, 2.86, 3.36, 3.86, 4.36, 4.86, 5.36, 5.86) = -4.00	0
10	min(-6.19, -0.65, -0.14, 0.36, 0.86, 1.36, 1.86, 2.36, 2.86, 3.36, 3.86) = -6.19	0

(15.33, 19.58, 19.04, 18.75, 18.84, 19.24, 19.89, 20.72, 21.64, 22.61, 23.60) and these correspond to ordering 0, 1, 2,..., 10 items. The figure for an order of 3 items (i.e. order up to an inventory level of 10 items) is a local minimum which is £0.09 less than ordering 4 items and £0.29 less than ordering 2 items.

Such figures give some sense of the sensitivity of the optimal policy to an error in the decision; with the data that have been given, it is less serious to have an overstock of one item than to understock by one item. Once the policy has been found, it is possible to identify the way that the expected cost is built up from the components of ordering costs, holding costs and disposal costs. This information is of some, though very limited, value in exploring the sensitivity of the solution to changes in the data and cost coefficients. Common sense allows one to state some heuristic rules for determining the sort of sensitivity behaviour that is to be expected, with the assumption that all costs remain fixed except one. If the order cost is increased, then one would expect that the optimal policy is to try and place fewer orders, which would lead to an increase in the target

Table 6.5: Returns and orders (week 2)

s	$f_3^*(s)$ in £s	Optimal order
0	min(32.32, 31.39, 30.39, 29.68, 29.41, 29.52) = 29.41	9
1	min(29.82, 28.89, 27.89, 27.18, 26.91, 27.02, 27.45) = 26.91	8
2	min(27.32, 26.39, 25.39, 24.68, 24.41, 24.52, 24.95, 25.57) = 24.41	7
3	min(24.82, 23.89, 22.89, 22.18, 21.91, 22.02, 22.45, 23.07, 23.78) = 21.91	6
4	min(22.32, 21.39, 20.39, 19.68, 19.41, 19.52, 19.95, 20.57, 21.28, 22.02) = 19.41	5
5	min(14.82, 18.89, 17.89, 17.18, 16.91, 17.02, 17.45, 18.07, 18.78, 19.52, 20.27) = 14.82	0
6	min(11.39, 15.39, 14.68, 14.41, 14.52, 14.95, 15.57, 16.28, 17.02, 17.77, 18.52) = 11.39	0
7	min(7.89, 12.18, 11.91, 12.02, 12.45, 13.07, 13.78, 14.52, 15.27, 16.02, 16.77) = 7.89	0
8	min(4.68, 9.41, 9.52, 9.95, 10.57, 11.28, 12.02, 12.77, 13.52, 14.27, 15.02) = 4.68	0
9	min(1.91, 7.02, 7.45, 8.07, 8.78, 9.52, 10.27, 11.02, 11.77, 12.52, 13.27) = 1.91	0
10	min(-0.48, 4.95, 5.57, 6.28, 7.02, 7.77, 8.52, 9.27, 10.02, 10.77, 11.52) = -0.48	0

inventory at some stages. So it proves with the data given: if the order cost is increased to £6.00 (a 20% increase) then the changed optimal policy for $f_4^*(s)$ is in Table 6.7.

The difference between the original $f_4^*(s)$ and those in Table 6.7 arise from the extra £1.00 for the order which must be placed, the extra holding cost which will be expected and the risk of having to place a further order, which is now less likely than before. Another change which might be of interest would be the consequence of an increased holding cost. Common sense would lead one to expect that this would tend to reduce the target inventory and increase the expected costs of optimal policies. Table 6.8 shows the consequences of doubling the holding cost per item per week to £0.50 on $f_1^*(s)$:

The target inventory at stage 1 has fallen to 9 items; the cost of this optimal policy is only £0.05 more than the policy which would order up to 8 items.

Table 6.6: Returns and orders (week 1)

s	$f_4^*(s)$ in £s	Optimal order
0	min(39.15, 38.59, 37.83, 37.08, 36.54, 36.25) = 36.25	10
1	min(36.65, 36.09, 35.33, 34.58, 34.04, 33.75, 33.84) = 33.75	9
2	min(34.15, 33.59, 32.83, 32.08, 31.54, 31.25, 31.34, 31.74) = 31.25	8
3	min(31.65, 31.09, 30.33, 29.58, 29.04, 28.75, 28.84, 29.24, 29.89) = 28.75	7
4	min(29.15, 28.59, 27.83, 27.08, 26.54, 26.25, 26.34, 26.74, 27.39, 28.22) = 26.25	6
5	min(21.65, 26.09, 25.33, 24.58, 24.04, 23.75, 23.84, 24.24, 24.89, 25.72, 26.64) = 21.65	0
6	min(18.59, 22.83, 22.08, 21.54, 21.25, 21.34, 21.74, 22.39, 23.22, 24.14, 25.11) = 18.59	0
7	min(15.33, 19.58, 19.04, 18.75, 18.84, 19.24, 19.89, 20.72, 21.64, 22.61, 23.60) = 15.33	0
8	min(12.08, 16.54, 16.25, 16.34, 16.74, 17.39, 18.22, 19.14, 20.11, 21.10, 22.10) = 12.08	0
9	min(9.04, 13.75, 13.84, 14.24, 14.89, 15.72, 16.64, 17.61, 18.60, 19.60, 20.60) = 9.04	0
10	min(6.25, 11.34, 11.74, 12.39, 13.22, 14.14, 15.11, 16.10, 17.10, 18.10, 19.10) = 6.25	0

6.7 Exercises

(1) What is the optimal policy in the die-rolling game if the die is to be rolled **four** times? What would a fair entry fee be?

(2) Statisticians occasionally use icosahedral dice; these have 20 faces, marked with the ten digits 0-9 twice. All ten digits have the same chance of turning up. Suppose you are playing the die-rolling game using such dice; what should your optimal strategy be? What would be a fair entry fee for the game?

(3) In the solution of the die-rolling game, mention was made of the fact that there was no charge for continuing with the game and not stopping. What happens if such a charge is present? To be specific, what happens if charges of: (a) 20p; (b) 60p; (c) £1.00; are made for the privilege of continuing?

(4) Suppose that the rewards in the die-rolling game are changed so that the reward for score X_j on the jth roll is £X_j^2; what optimal strategy should be

Table 6.7: The effect of changing the order cost by +20% on $f_4^*()$

s	$f_1^*(s)$ in £s	Optimal order
0	min(41.37, 40.72, 39.82, 38.90, 38.16, 37.68, 37.61, 37.89) = 37.61	11
1	min(38.87, 38.22, 37.32, 36.40, 35.66, 35.18, 35.11, 35.39) = 35.11	10
2	min(36.37, 35.72, 34.82, 33.90, 33.16, 32.68, 32.61, 32.89) = 32.61	9
3	min(33.87, 33.22, 32.32, 31.40, 30.66, 30.18, 30.11, 30.39, 30.97) = 30.11	8
4	min(31.37, 30.72, 29.82, 28.90, 28.16, 27.68, 27.61, 27.89, 28.47, 29.25) = 27.61	7
5	min(22.87, 28.22, 27.32, 26.40, 25.66, 25.18, 25.11, 25.39, 25.97, 26.75, 27.65) = 22.87	0
6	min(19.72, 24.82, 23.90, 23.16, 22.68, 22.61, 22.89, 23.47, 24.25, 25.15, 26.12) = 19.72	0
7	min(16.32, 21.40, 20.66, 20.18, 20.11, 20.39, 20.97, 21.75, 22.65, 23.62, 24.61) = 16.32	0
8	min(12.90, 18.16, 17.68, 17.61, 17.89, 18.47, 19.25, 20.15, 21.12, 22.11, 23.10) = 12.90	0
9	min(9.66, 15.18, 15.11, 15.39, 15.97, 16.75, 17.65, 18.62, 19.61, 20.60, 21.60) = 9.66	0
10	min(6.68, 12.61, 12.89, 13.47, 14.25, 15.15, 16.12, 17.11, 18.10, 19.10, 20.10) = 6.68	0

adopted and how much do you expect to win?

(5) You want to sell your car within the next three days; based on previous experience of selling cars privately, you expect the offers to arrive at the rate of one each day, each one independent of the others, and each one uniformly distributed over the range £1000 to £1200. You must decide on whether or not to accept an offer immediately and want to maximise your expected gain; how much should you expect to receive?

(6) In order to find an actor to play "Hamlet", a theatre producer plans to audition up to three men. She knows that actors' skills for the part are randomly distributed, with 20% having skills of rank 10 (the best), 30% skills of rank 9, 40% skills of rank 8, and the remaining 10% skills of rank 7. Because of the commitments of those she is to interview, she must make immediate decisions about whether to engage an actor immediately after an audition, and wants to

Table 6.8: The effect of doubling the holding cost on $f_4^*()$

s	$f_1^*(s)$ in £s	Optimal order
0	min(43.74, 43.15, 42.44, 41.95, 41.90, 42.26) = 41.90	9
1	min(41.24, 40.65, 39.94, 39.45, 39.40, 39.76, 40.66) = 39.40	8
2	min(38.74, 38.15, 37.44, 36.95, 36.90, 37.26, 38.16, 39.46) = 36.90	7
3	min(36.24, 35.65, 34.94, 34.45, 34.40, 34.76, 35.66, 36.96, 38.56) = 34.40	6
4	min(33.74, 33.15, 32.44, 31.95, 31.90, 32.26, 33.16, 34.46, 36.06, 37.86) = 31.90	5
5	min(26.24, 30.65, 29.94, 29.45, 29.40, 29.76, 30.66, 31.96, 33.56, 35.36, 37.27) = 26.24	0
6	min(23.15, 27.44, 26.95, 26.90, 27.26, 28.16, 29.46, 31.06, 32.86, 34.77, 36.74) = 23.15	0
7	min(19.94, 24.45, 24.40, 24.76, 25.66, 26.96, 28.56, 30.36, 32.27, 34.24, 36.23) = 19.94	0
8	min(16.95, 21.90, 22.26, 23.16, 24.46, 26.06, 27.86, 29.77, 31.74, 33.73, 35.73) = 16.95	0
9	min(14.40, 19.76, 20.66, 21.96, 23.56, 25.36, 27.27, 29.24, 31.23, 33.23, 35.23) = 14.40	0
10	min(12.26, 18.16, 19.46, 21.06, 22.86, 24.77, 26.74, 28.73, 30.73, 32.73, 34.73) = 12.26	0

maximise the expected ranking of the skills of the person employed. How should she proceed?

(7) A shop pays £45 to place an order for ceramic rallods plus £40 each. These are sold at £65 each. The demand for rallods per week is given in Table 6.9. Items not sold in a given week are held in store at a cost of £5 each per week. The shop is planning for a four-week period, after which the surplus stock may be returned for a credit of £40 per item, although it will cost £50 to administer the return of any goods. Using dynamic programming, Table 6.10 is obtained for the expected costs of an optimal policy given an inventory level of s at the start of week 2. Use this to calculate the expected cost of an optimal policy given opening inventories of a) 0 rallods and b) 5 rallods at the start of week 1.

(8) A marketing consultant approaches the shop described in exercise (7) and offers the manager the following business proposition: if the opening inventory, s, in any week is less than 5 rallods, then the consultant will provide an accurate

Table 6.9: Demand for rallods (exercise 7)

demand (k)	probability
0	0.000
1	0.125
2	0.375
3	0.375
4	0.125

Table 6.10: Partial results of dynamic programme (exercise 7)

s	$f_2^*(s)$ in £s	Optimal order
0	min(-6.72, -5.25, -10.94, -18.05, -21.37, -21.54, -15.78, -3.11, 12.23, 27.48) = -21.54	9
1	min(-46.72, -45.25, -50.94, -58.05, -61.37, -61.54, -55.78, -43.11, -27.77, -12.52) = -61.54	8
2	min(-86.72, -85.25, -90.94, -98.05, -101.37, -101.54, -95.78, -83.11, -67.77, -52.52) = -101.54	7
3	min(-126.72, -125.25, -130.94, -138.05, -141.37, -141.54, -135.78, -123.11, -107.77, -92.52) = -141.54	6
4	min(-211.72, -165.25, -170.94, -178.05, -181.37, -181.54, -175.78, -163.11, -147.77, -132.52) = -211.72	0
5	min(-250.25, -210.94, -218.05, -221.37, -221.54, -215.78, -203.11, -187.77, -172.52) = -250.25	0
6	min(-295.94, -258.05, -261.37, -261.54, -255.78, -243.11, -227.77, -212.52) = -295.94	0
7	min(-343.05, -301.37, -301.54, -295.78, -283.11, -267.77, -252.52) = -343.05	0
8	min(-386.37, -341.54, -335.78, -323.11, -307.77, -292.52) = -386.37	0
9	min(-426.54, -375.78, -363.11, -347.77, -332.52) = -426.54	0
10	min(-460.78, -403.11, -387.77, -372.52) = -460.78	0

forecast of the demand in that week for a fixed fee of £C. Then the shop will only need to place an order for rallods if the consultant's forecast is for a demand greater than s. How may dynamic programming be used to determine whether this proposition should be accepted?

(9) Sometime within the next 3 weeks, you intend to purchase a "freezer-pack" of meat from a wholesale butcher. Prices of meat fluctuate randomly, but the consumer group in your locality has plotted the prices for the last 100 weeks and thus estimated the probability of various prices being charged as those in Table 6.11.

Table 6.11: Meat prices and probabilities (exercise 9)

Price	£28	£29	£30	£31	£32	£33	£34
Probability	0.01	0.04	0.24	0.42	0.24	0.04	0.01

Assuming that there is no correlation between the prices in successive weeks, what strategy should you adopt in order to purchase the meat at minimal expected cost to you?

(10) Consider a simple inventory problem where one wishes to plan the buying strategy for two time periods only. The demand in each one is random, and sales of 1, 2, 3, 4 or 5 items are equally likely in the two time periods. It costs £10 to place an order for goods; there is a holding charge of £2 per item per period and an excess charge of £8 per item is charged if an item is required when the inventory level is zero. The profit per item sold is £8 and the loss made on items which are on hand at the end of the second period is £4. What is the optimal purchase quantity for opening inventory levels of: a) 1 item? b) 5 items?

7

Further Stochastic Models

7.1 Stochastic models with an uncertain duration

One characteristic of the dynamic programming models which have been examined so far is that the number of stages has been known in advance, either as a fixed number, or as a fixed upper limit. There are some cases where this need not be the case. A system may continue, with a decision being followed by an action repeatedly until a specified goal has been achieved. An everyday example is of a search. When one is looking for a lost object, a decision is made to look for it in a particular region for a given time. Then, if the search has been unsuccessful, a further choice of a region is made, and this continues until the object is found ... or the search is abandoned.

Similar problems occur in several areas of management decision making. In this chapter, we shall consider some demonstrations of them and see how to solve the problems when they are arranged as dynamic programming problems. We shall see how lessons from earlier formulations have their natural counterparts in the problems with uncertain durations.

7.2 A production control problem

Some manufacturing processes are not 100% reliable. There may be a production plan for 10 items; when the finished items are examined, some may be perfect, some may have minor faults which can be easily corrected, but some of will have flaws which make them worthless or of second-class quality. (In many towns in Britain, such goods are sold in flourishing shops which sell "Manufacturers' rejects" or "Seconds". Similar factory outlets are to be found in North America.) Knowing that this happens, many production plans make allowance for such unreliability in the process and produce excess items. Problems arise when these extra items have a small value compared with the costs involved in making them, or when the capacity of the manufacturing system is limited. Then the manufacturer has to decide how many items should be in a production batch (and some of these may prove to be second-class) in order to keep his costs within limits and to satisfy all the constraints of demand and capacity. In such circumstances, there is an obvious sequence of decisions over time. After each production run, the manager knows how many perfect items are still needed, and must decide how many to make in the next production run, should this be needed. The option of improving the quality of the manufacturing process is one that the operational research analyst ought to consider; we assume that this has been recognized.

We will assume that the manufacturer has an order for N items. A production run is limited to a maximum size of Q items, and a production run for quantity m will cost $\phi(m)$; this cost includes an element for the setup of a run, and costs that depend on the number of items produced. If the decision taken is to make m items, then with probability p_{km}, k good items will be produced $(0 \leq k \leq m)$. Good items are sold for C_g each; any which are surplus to the order can be sold for C_s each; rejected items are sold at C_r each. These unit costs are such that it is never worthwhile for a deliberate decision to be taken to overproduce the finished articles. C_r may be zero.

Suppose that the manufacturer has taken n decisions, and there are currently s good items available. If $s \geq N$, then there is no problem. But if not, then the manufacturer must decide on a production quantity m for the next manufacturing run. This decision will certainly cost $\phi(m)$; in return, there will be income from sales of finished items, and this income will be a random variable with two or three components. With probability p_{km} he will receive $(m-k)C_r$ from sale of the rejected items; if $k \leq (N-s)$ he will receive kC_g from the sale of good items with the same probability p_{km}; if $k > (N-s)$, with the same probability again, he will receive $(N-s)C_g + (k-N+s)C_s$ from the sale of good items and surplus items. Thus the expected profit (income from sales - costs) from this decision will be

$$\sum_{k=0}^{N-s} (p_{km}((m-k)C_r + kC_g))$$

$$+ \sum_{k=N-s+1}^{m} (p_{km}((N-s)C_g + (k-N+s)C_s + (m-k)C_r)) - \phi(m) \quad (7.01)$$

(if $m > (N-s)$) and:

$$\sum_{k=0}^{m}(p_{km}((m-k)C_r + kC_g)) - \phi(m) \tag{7.02}$$

(if not).

Under dynamic programming assumptions, this expression will form the single stage benefit from a decision; using the principle of optimality, we must follow an optimal policy from the state which results from the decision. So let us define $f_n^*(s)$ as the profit that is expected from an optimal policy, after n decisions, with the state defined by s good items. "Optimal" means that we want to maximise this function. Using the expressions above, we can define:

$$\begin{aligned} f_n^*(s) = \max\Bigg[& \max_{0\le m\le N-s}\left[\sum_{k=0}^{m}(p_{km}((m-k)C_r + kC_g + f_{n-1}^*(s+k))) - \phi(m)\right], \\ & \max_{N-s+1\le m\le Q}\Bigg[\sum_{k=0}^{N-s}(p_{km}((m-k)C_r + kC_g + f_{n-1}^*(s+k))) \\ & + \sum_{k=N-s+1}^{m}(p_{km}((N-s)C_g + (k-N+s)C_s)) \\ & + \sum_{k=N-s+1}^{m}(p_{km}((m-k)C_r + f_{n-1}^*(s+k))) - \phi(m)\Bigg]\Bigg] \end{aligned} \tag{7.03}$$

There will be "boundary conditions" that

$$f_n^*(s) = 0 \qquad \forall s \ge N$$

This formulation has one major flaw: there is no indication of the number of stages which will be needed to complete the order. Thus it is not possible to apply the normal rules of dynamic programming and work backwards from the last decision to the first. But this is not as serious a problem as it might seem; it doesn't matter what stage the process has reached; the decision about production quantities will always be the same for a given state; the suffix n can be removed from f^*.

If we do this, then it is possible to rearrange the expression. On the right hand side, $f^*(s)$ occurs in each of the alternative expressions, multiplied by p_{0m}. If we moved these terms to the left hand side, the expression would have $f^*(s)$ multiplied by $(1-p_{0m})$ and so one can revise the expressions to make $f^*(s)$ the subject yielding:

$$\begin{aligned} f^*(s) = & \\ \max\Bigg[& \max_{0\le m\le Ns}\left[\frac{(p_{0m}mC_r + \sum_{k=1}^{m}(p_{km}((m-k)C_r + kC_g + f^*(s+k)) - \phi(m))}{(1-p_{0m})}\right], \\ & \max_{Ns+1\le m\le Q}\Bigg[\frac{(p_{0m}mC_r + \sum_{k=1}^{Ns}(p_{km}((m-k)C_r + kC_g + f^*(s+k)))}{(1-p_{0m})} \\ & + \frac{\sum_{k=Ns+1}^{m}(p_{km}((Ns)C_g + (k-Ns)C_s + (m-k)C_r))) - \phi(m))}{(1-p_{0m})}\Bigg]\Bigg] \end{aligned} \tag{7.04}$$

where $Ns = N - s$

The solution process is now an iterative one; the right hand side involves $f^*(s+1)$, $f^*(s+2)$ and so on and gives an expression for $f^*(s)$. Thus, starting with the terms that are known and calculating the others in sequence, we can find all values of $f^*(s)$. In practice, we start with $f^*(N)$ and use this to find $f^*(N-1)$. Then we find $f^*(N-2)$, $f^*(N-3)$, ..., $f^*(0)$

7.3 Example of production planning

A factory has an order for 8 identical ceramic items. These will be sold for £100.00 each. Up to 10 items can be made in one production run, and there is a probability of 0.6 that any one item from a production run will be good. It costs £30.00 to set up a production run and each item in the run costs £20.00; items which are surplus have no value; nor do scrap items. What is the optimal production policy?

We know that $f^*(s) = 0$ for $s \geq 8$. Then, if we consider the possible decisions when $s = 7$, we find: †

$$\begin{aligned} f^*(7) = \max(&0.4f^*(7) + 0.6f^*(8) + 0.6 \times 100 - 30 - 20, && \text{make 1} \\ &0.16f^*(7) + 0.48f^*(8) + 0.36f^*(9) + 0.84 \times 100 - 30 - 40, && \text{make 2} \\ &0.064f^*(7) + 0.288f^*(8) + 0.432f^*(9) + 0.216f^*(10) \\ &\quad + 0.936 \times 100 - 30 - 60) && \text{make 3} \\ &\quad \text{etc.} \end{aligned}$$

(It is not necessary to consider any further decisions, because the income will always be less than the cost of production.)

If we perform the necessary calculations, and substitute for the expressions that are known, we have:

$$\begin{aligned} f^*(7) = \max(&0.4f^*(7) + 10, && \text{make 1} \\ &0.16f^*(7) + 14, && \text{make 2} \\ &0.064f^*(7) + 3.6) && \text{make 3} \end{aligned}$$

This can be rearranged and simplified to give:

$$\begin{aligned} f^*(7) = \max(&10/0.6 = 16.6667, && \text{make 1} \\ &14/0.84 = 16.6667, && \text{make 2} \\ &3.6/0.936 = 3.8462) && \text{make 3} \end{aligned}$$

† Instead of using the full mathematical formulation, this solution is expressed in a more obvious format without the numerous (and possibly confusing!) options that appeared in the expressions in the earlier section. Once you have worked through the example, it may be useful to read through the full formulation again.

Hence, the optimal decision is to produce either one item or two, giving $f^*(7) = 16.6667$. We are now in a position to calculate $f^*(6)$. Using the same formulation, we find that:

$$
\begin{aligned}
f^*(6) = \max(&0.4f^*(6) + 0.6f^*(7) + 0.6 \times 100 - 30 - 20, && \text{make 1}\\
&0.16f^*(6) + 0.48f^*(7) + 0.36f^*(8)\\
&+ 0.48 \times 100 + 0.36 \times 200 - 30 - 40, && \text{make 2}\\
&0.064f^*(6) + 0.288f^*(7) + 0.432f^*(8) + 0.216f^*(9)\\
&+ 0.288 \times 100 + 0.648 \times 200 - 30 - 60, && \text{make 3}\\
&0.0256f^*(6) + 0.1536f^*(7) + 0.3456f^*(8) + 0.3456f^*(9)\\
&+ 0.1296f^*(10) + 0.1536 \times 100 + 0.8208 \times 200 - 30 - 80, && \text{make 4}\\
&0.01024f^*(6) + 0.0768f^*(7) + 0.2304f^*(8) + 0.3456f^*(9)\\
&+ 0.2592f^*(10) + 0.07776f^*(11) + 0.0768 \times 100 + 0.91296 \times 200\\
&- 30 - 100, && \text{make 5}\\
&0.004096f^*(6) + 0.036864f^*(7) + 0.13824f^*(8) + 0.27648f^*(9)\\
&+ 0.311040f^*(10) + 0.186624f^*(11) + 0.046656f^*(12)\\
&+ 0.036864 \times 100 + 0.95904 \times 200 - 30 - 120) && \text{make 6}\\
&\text{etc.}
\end{aligned}
$$

and after rearrangement and simplification, this becomes:

$$
\begin{aligned}
f^*(6) = \max(&20/0.6 = 33.3333, && \text{make 1}\\
&58/0.84 = 69.0476, && \text{make 2}\\
&73.2/0.936 = 78.2051, && \text{make 3}\\
&72.08/0.9744 = 73.9737, && \text{make 4}\\
&61.552/0.98976 = 62.1888, && \text{make 5}\\
&46.1088/0.995904 = 46.2984) && \text{make 6}
\end{aligned}
$$

Clearly, the optimal decision is to make three items.

The method proceeds in the same way for successive iterations (and a computer program is extremely useful for this: manual calculation is somewhat tedious!) yielding values for $f^*(s)$ as in Table 7.1.

The figures for optimal profit ought to be compared with those which would be expected if the production process was completely efficient. Then the profit from state $s < 8$ would be $80(8-s)-30$, which will exceed the data in the second column above. The difference is a penalty for an unreliable process.

It will be noticed that the difference between successive values of $f^*(s)$ is nearly constant and increases as s decreases until the last lines of the table. The principal reason that these lines show a change is the limit on the amount that can be produced in one run. Without such a constraint, one would find that the

Table 7.1: Production of ceramics

State ($= s$)	Optimal Profit ($= f^*(s)$)	Decision
7	16.6667	1 or 2
6	78.2051	3
5	139.8058	5
4	203.4125	6
3	267.2479	8
2	331.7586	9
1	395.5276	10
0	454.2286	10

differences $f^*(s) - f^*(s+1)$ increased as s decreased, reflecting the economies of larger runs.

7.4 Sensitivity analysis of planning policy

In a problem such as this, there are numerous variables to be considered when one is interested in sensitivity analysis. The optimal strategy will depend, directly or indirectly, on all of them. Nonetheless, it is possible to make some general remarks about the sensitivity of the solution to the dynamic programming problem to variations in the parameters.

The cost structure of the production process will affect the choice of production run and the expected cost of the policy. If the fixed cost is increased, then in general, production runs will become larger to share the cost over more items. The converse applies as well. Similarly, if the cost per item increases, then production runs will become smaller; the manufacturer will prefer to make small production runs and not have too great a risk of overproducing.

7.5 Problems of search

In the opening section of this chapter, there was a passing reference to problems of locating a lost object. We now return to this type of problem, which is characterized by being one without a predetermined number of stages. It is easy to describe the type of practical setting in which search problems are relevant; an item has to be found, and it is known to lie in one of a limited number of areas. The user has some device for attempting to locate the item which can be used in one area at a time, and which requires some time period to complete a search in a target area, but this is imperfect and may fail to recognize the item in an area, even if it is actually present. The objective is to try and find the lost item, either as quickly as possible, or to maximise the chance that it is found within a fixed length of time.

(There is no reason why one should restrict consideration to looking for lost objects. Problems of this nature can be considered by telephone sales staff, trying to sell their products to customers. In that case, there is a chance that the "target" is available, and an expected benefit from a successful call. Similarly, oil and mineral exploration staff must make decisions about where to look, not for a "lost" object, but for a source of revenue for their company. In such cases, there is a cost associated with the search, which may vary between search sites, together with a random fluctuation in the rewards which may be obtained from a search.)

These are two separate problems; while the problem of maximising the chance of finding a lost object within a time limit can be posed as a stochastic problem with a fixed maximum number of stages, that of finding it as quickly as possible could possibly have an infinite number of stages (but, to be realistic, no search would ever last so long!) It is this unbounded problem that will occupy the discussion in this section. There are numerous parameters which could be included in the description of even the simplest search problem. Let us suppose for simplicity that we are searching in one of N regions, and the best estimate at present of the probability that the item is in region i is p_i (for $i = 1, \ldots, N$). If we spend one unit of time searching in the ith region, then with probability q_i we will find the object if it is present (and clearly we won't find it if it isn't there!). After the search has been completed, our estimates of p_i will alter on the basis of the added information that the search has provided.

We want to minimise the expected time that it takes to find the lost object. This makes a sensible function to incorporate in the objective function of a dynamic programme. The simplification of dividing the time-scale into units of search provides a convenient set of stages, at each of which a decision ("Where should the next search be made?") has to be considered. The state of our knowledge at the stages is best summarized by the vector of probabilities $p_1, p_2, \ldots, p_N$.

Suppose that the function $f^*(p_1, p_2, \ldots, p_N)$ represents the expected number of searches (or time units) needed to find the object, given that $p_1, p_2, \ldots, p_N$ is an estimate of the probabilities that the item is in each of the N areas. Suppose that the decision that we take is to search area j. Then there are two possible outcomes. Either we succeed in the search, and this has probability $p_j q_j$, or we fail, with probability $1 - p_j q_j$. If we succeed, then the duration of the search will be one time unit. If we fail, then the expected duration of the search will be $1 + f^*(\pi_1, \pi_2, \ldots, \pi_N)$ time units, where the vector $\pi_1, \pi_2, \ldots, \pi_N$ represents our revised estimate of the probabilities that the object is in each of the regions.

The choice of a region j will be made so as to minimise the expected length of the search, so that our recurrence relationship will be:

$$f^*(p_1,\ldots,p_N) = \min_{1\leq j\leq N}(p_j q_j + (1-p_j q_j)(1+f^*(\pi_1,\ldots,\pi_N)) \qquad (7.05)$$

All that remains is to devise an expression for the vector $\pi_1, \pi_2, \ldots, \pi_N$ in terms of the information that a failed search in region j has given. For this we have to turn to Bayes' theorem, or, equivalently, consider the link between prior and posterior probabilities. The probability π_i represents the probability that the item is in the ith region given an unsuccessful search in the jth region when the state was described by the vector $p_1, p_2, \ldots, p_N$. By Bayes' theorem, this can be represented as the ratio of:

probability(unsuccessful search in j **given** item in i)× probability(item in i)

to:

probability(unsuccessful search in j)

and the latter is equal to:

$$\sum_{k=1}^{N}(\text{probability(unsuccessful search in } j \textbf{ given} \text{ item in } k) \times \text{probability (item in } k)) \qquad (7.06)$$

The probabilities in the numerator can be replaced very easily (and then be incorporated into the denominator). When $i \neq j$, it is certain that a search in j will fail, so the probability of this is 1. The prior probability is p_i. If the search is unsuccessful in the actual where the item actually is, then we have an event with probability $(1-q_j)$ with prior probability p_j. This replacement yields:

$$\pi_i = \frac{p_i}{(1-p_j q_j)} \quad (i \neq j) \qquad (7.07)$$

$$\pi_j = \frac{(1-q_j)p_j}{(1-p_j q_j)} \qquad (7.08)$$

Stopping to consider the form of this relationship, it should make some kind of intuitive sense. Generally, the probability π_i is larger than p_i, which is logical because one would expect a failure in one region to improve the chance of finding the item in another. There are two exceptions: π_j is smaller than p_j because the search in j has been unsuccessful and so it is reasonable for one to conclude that the object is less likely to be there; π_i will be unchanged when either p_j or q_j is zero ... either the object is definitely not in area j or it could never be found if it were to be there!

Before we continue, it is worthwhile to think about the form of the recurrence relation. It expresses the function $f^*(p_1, p_2, \ldots, p_N)$ in terms of a multiple of the same function, evaluated at another point in N-dimensional space,

$(\pi_1, \pi_2, \ldots, \pi_N)$ (To be strictly accurate, the two vectors are both on a surface in the N-dimensional space, since the components are all non-negative and sum to unity.) One might think, if you are the sort of reader who thinks about the text while reading it, what advantage this relationship has. Unless you know the value of f^* everywhere, how can you calculate it anywhere? The answer is simply that we choose to find the optimal policy without being especially concerned about the value of f^*. The best choice of search region is the one which makes the product $p_j q_j$ as large as possible. This corresponds, in a rough way, to searching each time where the item is most likely to be found at that time. With that information, we can specify the best policy given the prior probabilities $(p_1, p_2, \ldots, p_N)$. (This rule also means that we never consider the exceptional case where either p_j or q_j took the value zero.) Repeated use of the rule after each search with the prior probabilities p_i replaced by the posterior probabilities means that we can find the optimal policy sequentially.

The optimal rule is a common-sense one. Its validity was demonstrated by Black (1965) who considered a more general problem, where there was a charge c_i for each search in the ith region. In this case, the optimal policy is to search in the region for which the expression $\frac{p_j q_j}{c_j}$ is largest (" ...shopping where you expect to get the most per dollar."). Where all the charges are the same, then this becomes the rule quoted in the last paragraph.

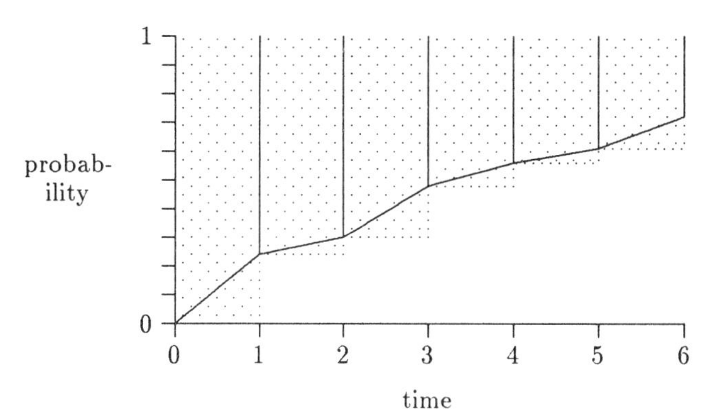

Figure 7.1: Success in sequential search

The argument which demonstrates the rule (ignoring any charge for searches) considers any policy which specifies where the first, second, third, ..., nth, ... searches should be made. Then if this policy is followed, there will be a probability, θ_k that the item is found on or before the kth search. (assume that $\theta_0 = 0$) The expected duration of the search can be written as:

$$\sum_{k=1}^{\infty}(1 - \theta_k) \tag{7.09}$$

(Black points out that this is a more reliable form for the expected duration than the more obvious $\sum_{k=1}^{\infty}(k(\theta_k - \theta_{k-1}))$ which ignores the possibility that the item is never found.) Then this expected duration can be pictured as the area shaded in Figure 7.1 where each vertical strip has width 1 unit and runs in height from θ_k to 1. The diagonal lines, joining the points (k, θ_k) to $(k+1, \theta_{k+1})$ divide the vertical strips into a triangle and a trapezium; the total of the areas of all the triangles will be constant and will not depend on the policy that is being followed. This is because on the kth search, you will be looking at some region j for the lth time ($l = 1, 2, 3, \ldots$), the probability of success is $p_j q_j (1 - q_j)^{l-1}$ and this does not depend on k. Each possible policy will correspond to some rearrangement of the searches, and so of the vertical strips of Figure 7.1. Clearly, the best thing to do is to follow the policy which makes the differences between successive θ-values as large as possible, making the diagonal lines decrease steadily in steepness. It follows immediately that this rearrangement is the one given by the rule quoted earlier. (Figure 7.1 does not show the optimal policy.)

7.6 Example of a search problem

A piece of jewellery has been lost in one of three places: on the pavement, by the canal bank or in a grassy field. At the outset, the three are equally likely. An hour spent searching along the pavement has probability 0.9 of finding the object; an hour by the canal bank has probability 0.6 of finding it; an hour in the field has probability 0.4 of success. How should we spend our time if we want to search most efficiently?

We can use the result quoted above directly; for the three areas being considered, the products of the prior probabilities and the chances of success are:

$$\begin{aligned}
\text{pavement} &= 0.9 \times 0.3333 = 0.30 \\
\text{canal bank} &= 0.6 \times 0.3333 = 0.20 \\
\text{grassy field} &= 0.4 \times 0.3333 = 0.1333
\end{aligned}$$

So the best place to spend the first hour searching is on the pavement. After the first hour, in our simplified model, we make a decision about where to go next, assuming that the jewellery has not yet been found. The first hour of work has altered the prior probabilities to be:

$$\begin{aligned}
\text{pavement} &= \frac{0.1 \times 0.3333}{(1 - 0.9 \times 0.3333)} = 0.0476 \\
\text{canal bank} &= \frac{0.3333}{(1 - 0.9 \times 0.3333)} = 0.4762 \\
\text{grassy field} &= \frac{0.3333}{(1 - 0.9 \times 0.3333)} = 0.4762
\end{aligned}$$

with the result that the product terms which are used to determine where to search next have values:

$$\begin{aligned} \text{pavement} &= 0.9 \times 0.0476 = 0.0429 \\ \text{canal bank} &= 0.6 \times 0.4762 = 0.2857 \\ \text{grassy field} &= 0.4 \times 0.4762 = 0.1905 \end{aligned}$$

Therefore the best place to spend the second hour searching will be on the canal bank. At the end of the second hour, again on the assumption that the jewellery has not been found, we have a further set of prior probabilities

$$\begin{aligned} \text{pavement} &= \frac{0.0476}{(1 - 0.6 \times 0.4762)} = 0.0667 \\ \text{canal bank} &= \frac{0.4 \times 0.4762}{(1 - 0.6 \times 0.4762)} = 0.2667 \\ \text{grassy field} &= \frac{0.4762}{(1 - 0.6 \times 0.4762)} = 0.6667 \end{aligned}$$

with the result that the product terms which are used to determine where to search next have values:

$$\begin{aligned} \text{pavement} &= 0.9 \times 0.0667 = 0.06 \\ \text{canal bank} &= 0.6 \times 0.2667 = 0.160 \\ \text{grassy field} &= 0.4 \times 0.6667 = 0.2667 \end{aligned}$$

In consequence the optimal plan is to spend the third hour searching in the field. Assuming that the first three searches are unsuccessful, one could continue and find the best policy for later searches. These are the areas listed in Table 7.2 in each time period.

Table 7.2: Optimal search policy

Search number	Area	Search number	Area	Search number	Area
1	pavement	2	canal bank	3	grassy field
4	canal bank	5	grassy field	6	grassy field
7	canal bank	8	pavement	9	grassy field
10	grassy field	11	canal bank	12	grassy field
13	grassy field	14	canal bank	15	grassy field

It is worth noting that the optimal policy includes periods where two successive hours are spent searching in the same area. One can also note that the pavement is not the best place to search again for several hours; this is a direct consequence of the high probability of successfully searching that area if the lost item is actually there; thus it would be unwise to return to the area again until other areas have been searched. (By extension, of course, if there were one or more areas in which a search would be certain to find the lost item if it were there, then those areas would only be searched once.)

7.7 Probabilities of success

The objective function has become secondary to the method of generating an optimal policy. It is worthwhile to return to the analysis of the expected number of searches which are needed to find the lost item. The objective function was defined as the minimal expected number of searches necessary; it is dependent on $2N-1$ variables ($N-1$ for the prior probabilities of success in the N regions – losing one variable because these probabilities must sum to unity – and N for the probabilities of success of searches). In general, the function will be of very little practical interest and the number of dimensions will make it almost impossible to express in a concise format. However, we can look at one aspect of its value using the optimal policy that has been used.

Suppose we define p_r^* as the probability of success during the rth search, given that the previous $r-1$ searches have failed. If the optimal policy is to search in areas $j_1^*, j_2^*, \ldots$ then it is straightforward to write down expressions for p_1^*, p_2^* and so on:

$$
\begin{aligned}
p_1^* &= p_{j_1^*} q_{j_1^*} \\
p_2^* &= p_{j_2^*} q_{j_2^*} \times (1 - p_1^*) \\
p_3^* &= p_{j_3^*} q_{j_3^*} \times (1 - p_1^* - p_2^*) \\
&\cdots\cdots \\
p_r^* &= p_{j_r^*} q_{j_r^*} \times (1 - \sum_{i=1}^{r-1} p_i^*)
\end{aligned}
$$

(using the rules for calculating the new prior probabilities after each of the searches to give the values of p_r.)

These expressions indicate that the optimal policy is the one which maximises the values of p_r^*, indicating a common sense explanation for the optimal policy, that you always search in the area where you are most likely to find the lost object. Using them, it would be possible to find the value of the objective function, as $\sum_{r=1}^{\infty}(rp_r^*)$. More usefully, you can express the probability of having found the object before a certain time has elapsed. This would bring the problem back to one of dynamic programming with a finite time horizon, and the optimal policy that has been used is one that will maximise the probability of success within any finite time period. (This accords, of course, with the principle of optimality from chapter 1.)

In the example of the search for the jewellery we can write down the first three probabilities of success as:

$$p_1^* = 0.9 \times 0.3333 = 0.30$$
$$p_2^* = 0.6 \times 0.4762 \times (1 - 0.30) = 0.20$$
$$p_3^* = 0.4 \times 0.6667 \times (1 - 0.30 - 0.20) = 0.1333$$

7.8 Sensitivity analysis in search problems

In many ways, the optimal choice of the first search area is very insensitive to the data. We determined this area by ranking the product terms $p_i q_i$ and the only circumstances in which the solution would be sensitive to variations in either the prior probabilities or the search probabilities would be if these variations led to a change in the ranking at the top of the list. So we can suggest the types of changes which would be significant and those which would not be significant.

Suppose that the optimal area to search first is area k and that area l is the second best, measured in terms of the product terms. (This means that the best area to search on the second search will be either area l or area k again.) Then we know that

$$p_k q_k > p_l q_l > p_i q_i \quad (\forall i \neq k, l)$$

If either p_k or q_k is increased, then k will still be the optimal area. If p_k is reduced, then one (or more) of the remaining prior probabilities must change to ensure that these terms sum to 1. Thus k will remain optimal only so long as the product term for k is the largest. If q_k is reduced, then this need not affect the other probabilities of success, so k will remain optimal as long as $q_k > p_l q_l / p_k$. Similar arguments can be derived to deal with changes to the other probabilities and their effect on the optimal search area.

7.9 Exercises

(1) You are required to produce 2 items using a process which is unreliable. A production run for this process can have size 1, 2 or 3 items, and will cost £25 plus £10 for each item included. The probability distribution of the number of good items produced is binomial with probability of success 0.75; such good items are sold for £100 each and the defective items are destroyed. Items that are surplus to production are also destroyed. What is the production strategy that maximises the expected profit to you?
Consider the optimal strategy when there is one good item as a function of the probability of success in the manufacturing process; at what values of this probability does the strategy change?

(2) The production process in a factory can make up to 4 items at a time. There is an order for 4 items, which will be sold for £200 each. A production run costs £50 plus £50 per item, with the probability distribution of successful items being binomial with probability of success 0.9. Defective items are destroyed and those that are surplus can be sold at a price of £C. At what values of C does the optimal strategy change?

(3) A manufacturer has an order for 4 items, and the production process can make up to 2 at a time. The production process costs £15 per item. If one item is made, then the probability that it is good is 0.9; if two are made, then with probability 0.6, both will be good and with probability 0.3, one will be good. Defective items are destroyed and surplus ones are sold at £50 each. Good items from the first three production runs are sold for £125 each, but from later runs the selling price falls to £100 each because of late delivery. What is the optimal production strategy?

(4) Your company has an order for 3 pieces of decorated pottery. You have a choice of two employees who can make such pieces. Alan's wages will be £25 per piece, and with probability 0.7, the result will be acceptable. Brian's wages will be £35 per piece, and with probability 0.8, the result will be acceptable. In each case, up to three pieces can be made at the same time. Formulate the problem of producing at least 3 pieces, at minimal expected cost, as a dynamic programme, and determine which employee should do the work.

(5) An item has been lost in one of two places, "here" and "there". The probability of finding it in either place is 0.36, but you consider that it is more likely to be "here" than "there". Your prior probabilities for the two areas are, respectively, $\frac{5}{9}$ and $\frac{4}{9}$. Show that the optimal policy for a search is to search the two areas alternately.

(6) "...when you have eliminated the impossible, whatever remains, however improbable, must be the truth." How does this quotation from Sherlock Holmes (in "The Sign of Four") have its parallel in the use of dynamic programming for search?

(7) Suppose that the prior probabilities for having lost the item of jewellery in the three areas of the example were changed to: pavement: $(1 - 2\alpha)$; canal bank: α; grassy field: α. What is the smallest value of α which would make you change your mind about the first area to be searched? With this new value, what sequence of areas would you search for the first four hours? And what would be the probability of success within that time?

(8) The Odeon Cinema in Woodbury is in a long straight street, which is marked out with parking spaces at equal intervals. Figure 7.2 illustrates these parking spaces, numbered from $-\infty$ to $+\infty$. There is thick fog on the evening when you wish to visit the cinema, so that you can determine

whether a parking space is empty or not, but cannot see the space ahead of you. Because of the fog, it would be dangerous to reverse to a space that you have driven past. Assuming that the probability of finding any space empty is p, formulate the problem of deciding when to park as a dynamic programme, whose objective is to minimise the expected distance that you and your partner will have to walk from the car to the cinema.

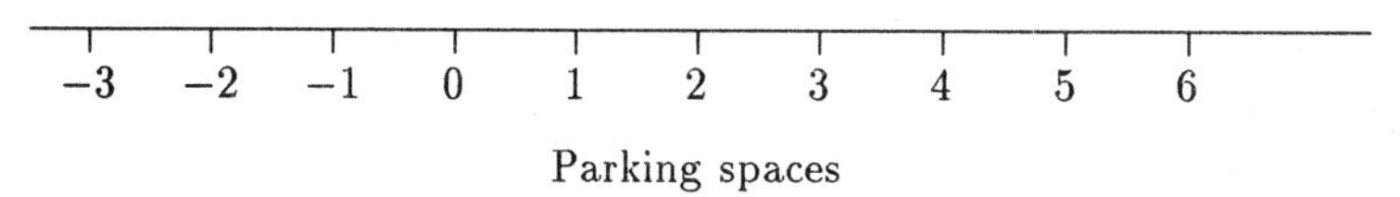

Figure 7.2: Parking on a foggy evening (exercise 8)

8

Stochastic Models with Infinite Duration

8.1 Introduction

In this chapter we shall look at other stochastic dynamic programming problems which extend the ideas that have been introduced in the last two chapters. Essentially the models to be introduced can be derived from the formulations that have been described already, but it is instructive to see the ways that the process of dynamic programming can be further extended.

8.2 Markov processes with gains

The model we shall consider first of all is an extension of the model used for stochastic inventory management. A decision maker has to choose a course of action at regular time intervals; when these decisions are taken, the state of the system is known, and can be described either quantitatively or qualitatively. (In the inventory control model, the state was described quantitatively.) Actions, too, may be measured numerically or simply identified as one from a list of possible options. By the time that the next decision time is reached, the system will have been transformed from its present state to a new one. The manner of this will be controlled by a random process which depends on the decision that has been taken.

So the state of the system follows a series of stochastic transitions, which may be summarized in a matrix of probabilities such as that in Figure 8.1 for one of the possible states. The rows of the matrix correspond to the decisions being taken and the columns correspond to the states that may be reached at the time when the next decision is to be taken. Alternatively, if the set of possible states of the system is the same at each decision point, one could draw up a matrix of transition probabilities for each of the decisions that exist (Figure 8.2). Such a matrix bears a strong similarity to the transition matrices of Markov processes.

$$A = \begin{pmatrix} 0.35 & 0.25 & 0.0 & 0.15 & 0.25 \\ 0.35 & 0.0 & 0.35 & 0.3 & 0.0 \\ 0.10 & 0.20 & 0.20 & 0.30 & 0.20 \end{pmatrix}$$

Figure 8.1: A transition matrix for a particular state and stage; three decisions, five possible future states

$$B = \begin{pmatrix} 0.8 & 0.2 & 0.0 & 0.0 & 0.0 \\ 0.35 & 0.0 & 0.35 & 0.3 & 0.0 \\ 0.15 & 0.15 & 0.2 & 0.3 & 0.2 \\ 0.0 & 0.2 & 0.25 & 0.3 & 0.25 \\ 0.0 & 0.35 & 0.2 & 0.45 & 0.0 \end{pmatrix}$$

Figure 8.2: A transition matrix for a particular decision and stage; five possible states

In addition to the transition probabilities, there are costs and benefits which are linked to both the decision being taken and the transition that follows. It may cost money to take a particular decision, and then there may be a reward (possibly random) as a result of the transition caused by the decision between states at successive stages. Looking back at the examples of inventory control will illustrate this; the decision to place an order involved the payment of an order cost and a cost per item; in the period before the next decision opportunity, random demand changed the state of the system, as measured by the inventory on hand, and this meant an income or reward for the manager of the stock.

With such a structured process, characterised by the repetition of the sequence:

Decision → Cost of decision → Transition → Benefit of transition

it is inevitable that it is sometimes referred to as a **Markov process with gains**. Besides inventory control examples, such processes arise in several commercial contexts. The examples used to illustrate the problems are, perhaps inevitably, very simplified, although they show some of the ideas which are involved in the formulation of dynamic programmes of the general kind.

8.3 Example of Markov process with gains

Consider a company which produces a range of quartz clocks, aimed at the gift-shop market. These are sold to wholesalers and most of the orders are placed at an annual trade fair. After each such trade fair, the sales staff analyse the comments of their buyers and their impressions of the trade fair and categorise the likely future demand for the existing range as being

i) high;
ii) medium;
iii) low.

The options available to the company at such times may be viewed as two binary ones, yielding four possibilities:

1) to do nothing;
2) to mail 5000 retail outlets with leaflets;
3) to hire an extra development engineer to come up with a new product in time for the next trade fair;
4) to do both 2) and 3)

There are costs associated with these options, and gains from the possible transitions which might occur as a result of the range of goods becoming more or less popular. We will suppose that the following data apply: for action 1), the transition matrix and returns (in thousands of pounds) are (ordered high, medium, low):

$$P_1 = \begin{pmatrix} 0.5 & 0.3 & 0.2 \\ 0.0 & 0.5 & 0.5 \\ 0.0 & 0.0 & 1.0 \end{pmatrix} \; R_1 = \begin{pmatrix} 60 & 50 & 40 \\ 50 & 40 & 30 \\ 40 & 30 & 20 \end{pmatrix}$$

for action 2), they are:

$$P_2 = \begin{pmatrix} 0.7 & 0.3 & 0.0 \\ 0.2 & 0.6 & 0.2 \\ 0.1 & 0.4 & 0.5 \end{pmatrix} \; R_2 = \begin{pmatrix} 55 & 45 & 35 \\ 45 & 35 & 25 \\ 35 & 25 & 15 \end{pmatrix}$$

for action 3) they are:

$$P_3 = \begin{pmatrix} 0.9 & 0.1 & 0.0 \\ 0.5 & 0.4 & 0.1 \\ 0.3 & 0.4 & 0.3 \end{pmatrix} \; R_3 = \begin{pmatrix} 45 & 35 & 25 \\ 35 & 25 & 15 \\ 25 & 15 & 5 \end{pmatrix}$$

and for action 4) they are:

$$P_4 = \begin{pmatrix} 1.0 & 0.0 & 0.0 \\ 0.6 & 0.4 & 0.0 \\ 0.4 & 0.5 & 0.1 \end{pmatrix} \; R_4 = \begin{pmatrix} 40 & 30 & 20 \\ 30 & 20 & 10 \\ 20 & 10 & 0 \end{pmatrix}$$

(The returns have been calculated assuming that it will cost £5000 for the mailing and £15000 for the recruitment; a high market will yield £60000 in a year, a medium one £40000 and a low one £20000; switches between two yield the mean of the appropriate figures. Thus, the entry of 25 (meaning £25000) in the second row and third column of matrix R_2 corresponds to a switch between a medium market (row 2) and a low market (column 3) with reward of $0.5 \times (40 + 20)$ minus the cost of a mailing (5). From the matrix P_2, this transition occurs with probability 0.2) In our subsequent discussion, we shall refer to the the entries in the matrix s^{th} row and t^{th} column of matrix P_d as $p_{d_{st}}$ and similarly for matrix R_d.

A problem such as this will readily fit into a dynamic programming formulation of the kind that we have already seen. The states have been identified already in the problem, and the stages are equally clear, as the successive times when the state can be measured and a decision taken. If the problem had a time horizon, possibly that the manufacturer wanted to cease production after a fixed number of years, then we would proceed as follows. Define the expected return $f_n^*(s)$ as being the return under an optimal policy from state s with n years (decisions) remaining. We assume that $f_0^*(s) = 0$ for all states. Then the recurrence relationship will be of the form:

$$f_n^*(s) = \max_d \left[\sum_t \left(p_{d_{st}} \left(r_{d_{st}} + f_{n-1}^*(t) \right) \right) \right] \tag{8.01}$$

where d represents the decision that is taken and t is an index of the possible states that might be reached at the next stage. This can be simplified by expressing the first term of the summation as one variable:

$$\sum_t p_{d_{st}} r_{d_{st}} = q_{ds} \tag{8.02}$$

and this may be regarded as the expected single-stage reward that follows from being in state s and taking decison d.

For the data that we have created, the matrix of these single-stage rewards will have three columns corresponding to the states and four rows for the decisions. It will be:

$$Q = \begin{pmatrix} 53 & 35 & 20 \\ 52 & 35 & 21 \\ 44 & 29 & 15 \\ 40 & 26 & 13 \end{pmatrix}$$

Here the best policy in the short term for a "high" market is to do nothing (expected reward 53), for one that is "medium"one should either do nothing or use a mail shot (expected reward 35), and for a "low" market, the best single stage return comes from the small expenditure on a mail shot (expected reward 21). The entries in row d and column s of this matrix are found by applying equation 8.2 to the four pairs of matrices P_d, R_d given earlier. This matrix Q

gives us the values of $f_1^*(s)$ directly and by use of the recurrence relationship, one could work back to $f_2^*(s)$ and so forth.

8.4 Howard's method

When faced with such an example, it is intuitively reasonable to expect that the optimal policy should settle to some rules which do not change, assuming that the planning horizon is far enough away. Policies for small values of n are likely to be affected by the 'running down' of the system towards the end of its life, but as that time gets further away, one expects that the best thing to do should depend on the state and not on the stage. This was seen in the inventory problems of chapter 6, where the policies tended to a limit of "order up to S items" as the time horizon became further away.

Given such an observation, we become interested in finding the policy which is optimal for large n. This problem was posed soon after the original definition of dynamic programming, and a solution was given by Howard []. He showed that in certain very general situations, the return from an optimal policy could be written as:

$$f_n^*(s) = v_s + ng \tag{8.03}$$

where v_s is the contribution to the return which results from being in state s and g is the average return per stage, taken over many stages. This result can then be used in an iterative solution procedure which takes a proposed policy as a trial solution, solves for the values of the unknown quantities in equation 8.01, and then uses these values to try and find a better policy. The two parts of this iterative procedure can be repeated until successive policies found by it are identical; then the process stops, with the last policy found as the optimal one. Generally, very few iterations will be needed.

The solution procedure starts by substituting for $f_n^*(s)$ and $f_{n-1}^*(s)$ from equation 8.03 in the recurrence relationship of 8.01; we assume that there is a possible policy, with a decision which has been suggested for each state. This will yield one linear equation in the vector $v_1, v_2, \ldots$ and g for each of the possible states. As a result, we have one less equation than there are variables; but this is not a difficulty, since one can solve them by setting one of the unknowns v_s to zero and solving for the remaining unknowns. The structure of the set of equations means that it is immaterial which of the unknowns becomes zero; the others will be calculated relative to it, and the differences between v_i and v_j will be the same irrespective of the term which is made equal to zero. With the answer to the set of simultaneous equations, referred to as the **value-determination algorithm**, one has numbers to feed in to the recurrence relationship to find the best policy for each state on the assumption that the values of $f_{n-1}^*(s)$ (on the right hand sides) follow equation 8.01. This will give a policy for each of the possible states and the process is referred to as the **policy improvement routine**. (It is easy to see that this will never give a worse policy than the one which was used before; the old policy provides a lower bound for the returns from the new one.) This yields a sequence shown in Figure 8.3.

Value-determination algorithm
Given a suggested policy for each state, solve for g and the set v_s, with one of the latter set to zero (usually the first or the last).

Policy improvement routine
Using the values found, try and find a better policy by substitution in the right-hand-side of the recurrence relationship.

Figure 8.3: Howard's method

8.5 Application of Howard's method

We are now in a position to solve for the optimal policy for the example of the clock manufacturer, using the two stages of the sequence in Howard's method. First we need a trial policy, and it seems reasonable to start with the policy which gives the best single stage returns, represented as a vector $(1, 1, 2)$

Using this policy in the expression of equation 8.01 we have the three equations:

$$\begin{aligned} v_1 + ng &= 0.5\,(60 + v_1 + (n-1)g) + 0.3\,(50 + v_2 + (n-1)g) \\ &\quad + 0.2\,(40 + v_3 + (n-1)g) \\ v_2 + ng &= 0.5\,(40 + v_2 + (n-1)g) + 0.5\,(30 + v_3 + (n-1)g) \\ v_3 + ng &= 0.1\,(35 + v_1 + (n-1)g) + 0.4\,(25 + v_2 + (n-1)g) \\ &\quad + 0.5\,(15 + v_3 + (n-1)g) \end{aligned} \tag{8.04}$$

which yields:

$$\begin{aligned} 0.5v_1 - 0.3v_2 - 0.2v_3 + g &= 53 \\ 0.5v_2 - 0.5v_3 + g &= 35 \\ -0.1v_1 - 0.4v_2 + 0.5v_3 + g &= 21 \end{aligned} \tag{8.05}$$

(three equations in four unknowns). Solving these by setting $v_3 = 0$ gives

$$v_1 = 51.6981 \quad v_2 = 9.8113 \quad v_3 = 0.0 \quad g = 30.0943 \tag{8.06}$$

(and it is worthwhile to check that the relative values of the v_s are not changed by altering the one which is set to zero). With these four values, we can enter the second phase, the policy improvement routine. This takes the four possible actions in each state and evaluates the sum of the expected single stage return and the expected benefit at the end of the transition, that is (for state s and decision d):

$$q_{ds} + \sum_t p_{d_{st}} v_t \tag{8.07}$$

and then finds the largest of these test quantities. Applied to our data, Table 8.1 shows the values obtained.

Table 8.1: Policy improvement routine (iteration 1)

state	decision	test quantity
1	1	81.7925
1	2	91.1321
1	3	91.5094
1	4	91.6981†
2	1	39.9057
2	2	51.2264
2	3	58.7736
2	4	60.9434†
3	1	20.0000
3	2	30.0943
3	3	34.4340
3	4	38.5849†

The new policy is to take decisions $(4, 4, 4)$, (whose returns are marked by the symbol † in the table) which is different from that tested first of all, so that we need to solve a new set of simultaneous equations:

$$\begin{aligned} g &= 40 \\ -0.6v_1 + 0.6v_2 + g &= 26 \\ -0.4v_1 - 0.5v_2 + 0.9v_3 &= 13 \end{aligned} \tag{8.08}$$

Solving these again by setting $v_3 = 0$ gives

$$v_1 = 42.9630 \quad v_2 = 19.6296 \quad v_3 = 0.0 \quad g = 40.0 \tag{8.09}$$

a better average gain per stage than before. So, for a second time, we need to use the policy improvement routine and the values in Table 8.2 show the results.

Table 8.2: Policy improvement routine (iteration 2)

state	decision	test quantity
1	1	80.3704
1	2	87.9630†
1	3	84.6296
1	4	82.9630
2	1	44.8148
2	2	55.3704
2	3	58.3333
2	4	59.6296†
3	1	20.0000
3	2	33.1481
3	3	35.7407
3	4	40.0000†

The new policy is to take decisions $(2, 4, 4)$, which is again a different one. For the third time, we need to solve simultaneous equations:

$$\begin{aligned} 0.3v_1 - 0.3v_2 + g &= 52 \\ -0.6v_1 + 0.6v_2 + g &= 26.0 \\ -0.4v_1 - 0.5v_2 + 0.9v_3 + g &= 13 \end{aligned} \tag{8.10}$$

which yield:

$$v_1 = 49.7531 \quad v_2 = 20.8642 \quad v_3 = 0.0 \quad g = 43.3333 \tag{8.11}$$

and the value of g has increased further. Using the policy improvement routine again we have Table 8.3.

The policy that appears from Table 8.3 is the same, that is $(2, 4, 4)$. The company should always use the mail shot of its products, and if the state of the market is judged to be anything other than high, the engineer should be hired. This will generate revenue of £43333 per annum in the long run.

8.6 Sensitivity analysis

When it comes to sensitivity analysis of this type of problem, we can consider two aspects of it. First, there is the same kind of sensitivity as has been encountered in earlier types of problem, wherein the solution that is obtained depends on the values that are taken by the parameters of the model. Second, the solution process depends on the initial approximation to the optimal policy.

Table 8.3: Policy improvement routine (iteration 3)

state	decision	test quantity
1	1	84.1358
1	2	93.0864†
1	3	90.8642
1	4	89.7531
2	1	45.4321
2	2	57.4691
2	3	62.2222
2	4	64.1975†
3	1	20.0000
3	2	34.3210
3	3	38.2716
3	4	43.3333†

The sensitivity of the solution to the model parameters follows the same general principles as have been observed in other dynamic programming models. The solution to the problem is in two parts; function values and policy decisions. Suppose that we change one of the expected gains from a decision. Then we can distinguish four cases, depending on whether the change is an increase or a decrease, and on whether the decision in question is in an optimal policy or not. Table 8.4 summarises these cases.

To what extent does the amount of calculation required in finding the optimal solution depend on the initially guessed solution? Even if our starting point had been the optimal solution, it would have been necessary to use the Value-Determination and the Policy-Improvement routines once each. If our starting point had been any other solution, then the two routines would have been used at least twice each. The point which was used in the example caused the routines to be used three times each; although the work required in these cases is not excessive, it would be helpful to have some idea about the dependence of the amount of work needed on the initial solution.

In the example quoted, there are 64 possible initial solutions of which all but seven converge to the optimal solution after two or three uses of the two stages of Howard's method. Of the seven, one is the optimal solution, leaving six initial solutions which never yield the optimal solution. These are:

$$(2,4,1), (3,4,1), (4,1,1,), (4,2,1), (4,3,1), (4,4,1)$$

and the common feature which they share is that all give reducible transition

Table 8.4: Sensitivity analysis for Markov processes with gains

	Decrease gain from decision	Increase gain from decision
Decision in an optimal policy	**Function value:** Decreases and then remains constant; **Policy decisions:** Remain unchanged until the gain falls so low that a new policy is optimal.	**Function value:** Increases; **Policy decisions:** Remain unchanged.
Decision not in an optimal policy.	**Function value:** Remains constant and then increases; **Policy decisions:** Remain unchanged until the gain reaches a value which introduces it to a new optimal policy.	**Function value:** Never changes; **Policy decisions:** Never change.

matrices; in other words, if the company management followed one of these six policies, it would be impossible to reach all states of the system from all other ones. For instance, the initial solution $(4, 1, 1)$ gives a transition matrix:

$$\begin{pmatrix} 1.0 & 0.0 & 0.0 \\ 0.0 & 0.5 & 0.5 \\ 0.0 & 0.0 & 1.0 \end{pmatrix}$$

Policy 1 in state 3 ensures that state 3 is perpetuated for ever; Policy 4 in state 1 retains that state for ever. The six possibilities for policies from states 1 and 2 never allow for the system to enter state 3. As a result, the evaluation routine cannot identify the relative values of the v_i terms.

These observations should serve as a warning when using Howard's method. Provided that the initial policy is chosen to give an irreducible transition matrix, then a solution will be found.

8.7 Discounting future values

It is common to use some kind of discounting when considering the financial implications of decisions whose effects will be observed over a period of several years. Underlying this idea is the recognition that a profit of £1000 now is worth more than the same amount one year into the future. The money can be used now for some investment in goods or services, or it can be invested;

in the latter case, one would expect that it will gain in value in the next year. The need for such a practice becomes more relevant when a decision-maker is contemplating a sequence of choices that lead to diverse outcomes. If you were offered an investment that paid you nothing now, £1000 in one years' time, the same amount in two years' time and £8000 in three years' time, would you prefer that to one which would cost you £2000 now, and pay you nothing in one year's time, £10000 in two years' time and a further £2000 in three years' time? Some business investments lead to similar irregular returns, and the problem may be further complicated by uncertainty so that the returns would be the expected values, and not deterministic ones. (The traditional english proverb "A bird in the hand is worth two in the bush" illustrates the idea; something which is known for certain, now, should be valued more highly than its equivalent which is known vaguely, and may not be obtained for certain.)

Because dynamic programming is concerned with decisions which are taken sequentially over time, there will be some situations in which income and expenditure of this nature has to be considered. The simplest way of representing the effect of inflation is to use a discounting factor. Thus £1 now will be equivalent to £$(1 + r)$ in one year's time, where r is the assumed rate of inflation; alternatively, £1 in one year's time, will correspond to £$\frac{1}{(1+r)}$ now. Most analysis works on the basis of expressing future costs and income in terms of the present value, so the fraction $\frac{1}{(1+r)}$ is often expressed as an annual discount factor α. (One assumes that this aspect, at least, of the future is known, and reasonably constant. This is not the place for a discussion about such an assumption!)

Discounting can then be incorporated into many of the models that we have developed in the book. In the discussion of the paths of minimal cost, we assumed that one unit of expenditure on the arcs at the end of the path was equivalent to one unit of expenditure at the beginning. This was reasonable as the assumption was that such paths will take very little time to traverse. However, if the time between successive decisions is significant, then discounting may be necessary. The recurrence relation for the problem of the stagecoach (Chapter 2) appeared in the form

$$f_n^*(i_n) = \min_{i_{n-1}}\{p(i_n, i_{n-1}) + f_{n-1}^*(i_{n-1})\} \tag{8.12}$$

where $p(i_n, i_{n-1})$ was the cost of travel from node i_n to node i_{n-1} and $f_n^*(i_n)$ was the cost of an optimal policy from stage n and state i_n. If the time taken to go from one stage to the next corresponded to a discount factor α then equation (8.12) would be replaced by

$$f_n^*(i_n) = \min_{i_{n-1}}\{p(i_n, i_{n-1}) + \alpha f_{n-1}^*(i_{n-1})\} \tag{8.13}$$

Here, any cost incurred after one stage is discounted by the factor α; because of the recurrence formula, costs incurred after two stages will be discounted by α^2, and so on. An example will clarify this situation, and introduce a possible area of application which offers scope for further dynamic programming modelling.

A company uses a fleet of similar vehicles; decisions about whether to replace them are taken annually on the anniversary of purchase, and repairs are carried out whenever necessary in between such decisions. Costs have been recorded for the expected repair costs for a year for vehicles of various ages, together with the expected resale value if a vehicle is to be sold. These appear in Tables 8.5 and 8.6.

Table 8.5: Expected repair costs

Age (in years) at start of year (y)	0	1	2	3	4
Annual cost of repair: $R(y)$	£2500	£3750	£5500	£8000	£10000

Table 8.6: Expected revenue from resale

Age of vehicle (years) (y)	1	2	3	4	5
Expected resale value $V(y)$	£20000	£15000	£11000	£7500	£6000

A new vehicle will cost £25000 and this price is expected to rise by 5% per year; to allow for the uncertainty of inflation and the possibility of changing the technology, a discount figure of $\alpha = 0.8$ is used; assuming that plans are to be made for the next seven years when should a vehicle that is new now be replaced?

With annual decisions, the state of the system is well-defined as the age of the current vehicle; stages are equally clearly identified as the times when decisions are made, possibly during a particular month of each year. And, as normal, we can work backwards from the final stage, the seventh year, when we shall expect to sell the current vehicle.

Suppose that we define $f_n^*(s)$ as the expected cost of an optimal policy with a vehicle of age s years and with n decisions (years) still remaining. Then, at stage $n = 0$, the best thing to do is to sell the vehicle, giving optimal objective function values derived directly from Table 8.6; $f_0^*(s) = -V(s)$ (the negative sign indicating that this is an income, not an expenditure).

The recurrence relationship between the values of the objective function at successive stages can be summarised as either:

the cost of selling the existing vehicle now, buying a new one and incurring the repair costs for one year, followed by the discounted value of the optimal policy of having a one-year-old vehicle at the next stage;

or:

the cost of keeping the existing vehicle and having to repair it for the next year followed by the discounted value of the optimal policy of having a vehicle that is one year older than currently at the next stage.

and we want to choose the cheaper option. Symbolically, this becomes:

$$f_n^*(s) = \min\Big(\big(-V(s) + (1.05)^{(7-n)} \times 25000 + R(0) + 0.8 f_{n-1}^*(0)\big)\,, \\ \big(R(s) + 0.8 f_{n-1}^*(s+1)\big)\Big)$$

So, for instance, $f_1^*(3) = \min((-11000 + 33502 + 2500 - 0.8 \times 20000) = 9002, (8000 - 0.8 \times 7500) = 2000) = 2000$ corresponding to the decision to retain the vehicle for the next year.

By working backwards, it is straightforward to tabulate the alternative values which contribute to $f_1^*(s)$ as in Table 8.7.

Table 8.7: Data for $f_1^*()$

State	Cost if sold	Cost if kept	Decision
0	∞	20002	Keep
1	2	-8250	Keep
2	5002	-3300	Keep
3	9002	2000	Keep
4	12502	5200	Keep
5	14002	∞	Sell

Similar tables can be drawn up for each of the earlier stages, and these are presented in Tables 8.8 and 8.9

The policy which emerges is that of keeping a new vehicle (expected cost £48297 in Table 8.9), sell it after two years (Table 8.9) and then retain the vehicle for a further five years. There is very little difference between the expected costs which determine $f_5^*(2)$ (£27808 or £27888) and so it is evident that this part of the optimal policy may be sensitive to some of the data given in the problem. The precise form of the sensitivity of the solution to each part of the data is hard to specify, although the general principles can be established qualitatively. Thus, with all other data fixed:

An increase in the cost of a new vehicle will tend to lead to a policy which retains vehicles until they are older;

Table 8.8: Data for $f_2^*()$, $f_3^*()$, $f_4^*()$
(optimal decision in bold)

State	$f_2^*()$ [sell,keep]	$f_3^*()$ [sell,keep]	$f_4^*()$ [sell,keep]
0	∞, **27807**	∞, **33776**	∞, 38985
1	7807, **1110**	13776, **9430**	18985, **15932**
2	12807, **7100**	18776, **15228**	23985, **23721**
3	16807, **12160**	**22776**, 24246	**27985**, 29021
4	**20307**, 21202	**26276**, 27446	**31485**, 32221
5	**21807**, ∞	**27776**, ∞	**32985**, ∞

Table 8.9: Data for $f_5^*()$, $f_6^*()$, $f_7^*()$
(optimal decision in bold)

State	$f_5^*()$ [sell,keep]	$f_6^*()$ [sell,keep]	$f_7^*()$ [sell,keep]
0	∞, **42808**	∞, **46931**	∞, **48927**
1	22808, **22726**	26931, **25997**	**28297**, 28507
2	**27808**, 27888	31931, **30947**	**33297**, 34245
3	**31808**, 33188	**35931**, 36247	**37297**, 39545
4	**35308**, 36388	**39431**, 39447	**40797**, 42745
5	**36808**, ∞	**40931**, ∞	**42297**, ∞

An increase in the discount rate will make future decisions less attractive financially.

8.8 Exercises

1) In order to help look after our garden, my wife and I employ a part-time gardener for a few hours each week; for this we pay either £25 or £30 and choosing the rate of pay is a decision that we have to make. Since the contract with the gardener is a casual one, we have found that sometimes the person employed decides to cease working for us. The probability that this happens depends on the amount being paid. Details are given in Table 8.10.
When a gardener leaves, we have found that we must do two things: first, we advertise for a replacement, either in a newspaper (£5 per week) or through an enployment agency (£15 per week); second, we must employ a company of landscape gardeners at a charge of £50 per week. Advertising, if successful, will produce a replacement gardener in the week after it has appeared, and the probability of success is shown in Table 8.10.

Table 8.10: Gardening in Exeter (exercise 1)

state (i)	action (k)	transition probabilities p_{i1k}	p_{i2k}	weekly cost £
1=gardener	low pay	0.85	0.15	25
	high pay	0.9	0.1	30
2=company	newspaper	0.4	0.6	55
	agency	0.7	0.3	65

What should be the rate of pay for our gardener? And how should we advertise for a replacement?

Table 8.11: Aceriman mineral cargoes (exercise 2)

port (i) (loading)	product (k)	transition probabilities p_{i1k}	p_{i2k}	p_{i3k}	p_{i4k}
1	1	0.00	0.10	0.30	0.60
	2	0.00	0.40	0.40	0.20
2	1	0.50	0.00	0.30	0.20
	2	0.30	0.00	0.50	0.20
3	1	0.30	0.40	0.00	0.30
	2	0.20	0.20	0.00	0.60
4	1	0.20	0.30	0.50	0.00
	2	0.40	0.60	0.00	0.00
		revenue r_{i1k}	r_{i2k}	r_{i3k}	r_{i4k}
1	1	0.00	38.00	30.00	34.00
	2	0.00	39.00	25.00	32.00
2	1	34.00	0.00	30.00	26.00
	2	38.00	0.00	31.00	24.00
3	1	32.00	36.00	0.00	40.00
	2	25.00	32.00	0.00	39.00
4	1	27.00	23.00	31.00	0.00
	2	32.00	25.00	39.00	0.00

2) A coastal cargo ship is used to move one of two mineral products between four ports in Acerima. The ship's owners must decide which product should be loaded in the ports in order to sign a contract for the next two years. They have the data of Table 8.11 relating to the probabilities that a load of product k, loaded in port i will be needed in port j (p_{ijk}) and the associated profit (r_{ijk}). The ship takes one week for each journey (loading, travel, unloading), so the contract may be regarded as being of infinite duration.

3) What is the optimal policy for a company which plans to operate a fleet of identical vehicles for the next six years with annual decisions about replacement given the data in Tables 8.12 and 8.13? (y is the age in years, $R(y)$ the annual cost of repair and $V(y)$ the expected resale value.) Assume that a new vehicle will cost £12000, with annual increases of 8%, and a discount figure of 0.8.

Table 8.12: Expected repair costs (exercise 3)

(y)	0	1	2	3	4	5
$R(y)$	£300	£750	£2000	£5000	£6000	£6000

Table 8.13: Expected revenue from resale (exercise 3)

(y)	1	2	3	4	5	6
$V(y)$	£16000	£12000	£9000	£7000	£5000	£4000

4) Suppose that you have been given the expected costs of repair ($R(y)$) for vehicles of ages $0, 1, 2, \ldots$, and the expected income from a sale ($V(y)$) for vehicles of ages $1, 2, \ldots$. The price of a new vehicle is £N and the price of a second hand vehicle is $1.2V(y)$. With annual price rises of p% and a discount figure of α, formulate the following decision problem: what is the optimal age of vehicle to purchase to minimise expected costs, and at what age should it be sold? How could this model be made more realistic?

9

Dynamic Programming for Fun

9.1 Apologia

This chapter introduces a more lighthearted aspect of dynamic programming. It is concerned with the use of dynamic programming for solving puzzles and games, not with models which might apply to commercial and industrial systems. It's here because the ideas which we shall look at are of interest in their own right, and the applications do have an educational value – or at least the author thinks that they do. So if you bought this book thinking that it would be 100% serious, this chapter is not for you. Some of the examples and exercises given earlier in the text have had their lighter aspects as well, but they were all in a more serious context.

9.2 A classic problem

You have been given two empty jugs; one holds 0.5 litres, the other holds 0.8 litres. Neither has any other markings on them, and you have no other measuring devices available. How can you measure exactly 0.7 litres of water into the larger jug with the two jugs? You are allowed to fill each one as often as you like, and can pour from one to the other if you want.

Obviously, by trial and error, it is possible to devise a solution to this problem. It would be of the form: "Fill the larger jug, transfer 0.5 litres to the smaller jug, discard ...". It doesn't take too long to work out the sort of transfers which could be made and to solve the puzzle. If you haven't met this question before, why not try it now?

In whatever way you decided to solve the problem, there will have been an implicit series of decisions. There are six actions which are possible in general:

1. To fill the 0.5 litre jug from a tap.
2. To fill the 0.8 litre jug from a tap.
3. To empty the contents of the 0.5 litre jug down a drain.
4. To empty the contents of the 0.8 litre jug down a drain.
5. To transfer water from the 0.5 litre jug to the 0.8 litre jug until either the small jug is empty or the large one is full.
6. To transfer water from the 0.8 litre jug to the 0.5 litre jug until either the small jug is full or the large one is empty.

Not all of these decisions are always possible or even particularly sensible. In general, the actions work to change the state of the system, described in terms of the contents of the two jugs. Such an interpretation of the process of looking for the answer points to a dynamic programme, with states and decisions appearing "naturally" from consideration of the problem. The stages follow naturally as well; these are those moments when one of the six decisions can be made. The decisions lead straight to a deterministic transition rule between states and stages; all that remains is an objective function. According to the traditional way that this problem is posed, the objective is to minimise the number of actions that you need to take before reaching the goal. So we'll look at this one first.

Define the objective $f^*\binom{e}{f}$ to be the minimal number of actions from a state defined by $\binom{e}{f}$ to achieve the objective. We can define e and f to be the number of multiples of 0.1 litre in the *eight* jug and the *five* jug respectively. The objective will be reached if we have one of the two states

$$\begin{pmatrix}7\\5\end{pmatrix}, \begin{pmatrix}7\\0\end{pmatrix}$$

The recurrence relation between the states is also fairly straightforward, although it looks a little complicated when we pose it in a general mathematical format. There are the six options described above and so:

$$\begin{aligned} f^*\begin{pmatrix}e\\f\end{pmatrix} = 1 + \min\Bigg[& f^*\begin{pmatrix}e\\5\end{pmatrix}, f^*\begin{pmatrix}8\\f\end{pmatrix}, f^*\begin{pmatrix}e\\0\end{pmatrix}, f^*\begin{pmatrix}0\\f\end{pmatrix}, \\ & \text{either } f^*\begin{pmatrix}8\\e+f-8\end{pmatrix} \text{ or } f^*\begin{pmatrix}e+f\\0\end{pmatrix}, \\ & \text{either } f^*\begin{pmatrix}e+f-5\\5\end{pmatrix} \text{ or } f^*\begin{pmatrix}0\\e+f\end{pmatrix}\Bigg] \end{aligned}$$

with each line corresponding to one of the options. This general formulation needs one further comment; the only possible states of the system are those in which at least one of the four terms $e, f, (8-e), (5-f)$ is zero, since these are the only ways that we would know when to stop filling or transferring liquid. Then at least one of the options will be unavailable because the corresponding action could not happen. Normally only two states on the right hand side of the expression will be different from that on the left hand side.

The easiest way to work through this problem is to go backwards from the two final states (where $f^*() = 0$) and find the states which could precede them, steadily progressing backwards to the initial state of the system where both jugs were empty. Thus we have:

$$f^*\begin{pmatrix}7\\5\end{pmatrix} = f^*\begin{pmatrix}7\\0\end{pmatrix} = 0$$

and therefore:

$$f^*\begin{pmatrix}8\\4\end{pmatrix} = f^*\begin{pmatrix}2\\5\end{pmatrix} = 1$$

The preceding states to these two yield:

$$f^*\begin{pmatrix}0\\4\end{pmatrix} = f^*\begin{pmatrix}2\\0\end{pmatrix} = 2$$

and the repeated process gives:

$$f^*\begin{pmatrix}4\\0\end{pmatrix} = f^*\begin{pmatrix}0\\2\end{pmatrix} = 3 \qquad f^*\begin{pmatrix}4\\5\end{pmatrix} = f^*\begin{pmatrix}8\\2\end{pmatrix} = 4$$

$$f^*\begin{pmatrix}8\\1\end{pmatrix} = f^*\begin{pmatrix}5\\5\end{pmatrix} = 5 \qquad f^*\begin{pmatrix}0\\1\end{pmatrix} = f^*\begin{pmatrix}5\\0\end{pmatrix} = 6$$

$$f^*\begin{pmatrix}1\\0\end{pmatrix} = f^*\begin{pmatrix}0\\5\end{pmatrix} = 7 \qquad f^*\begin{pmatrix}1\\5\end{pmatrix} = f^*\begin{pmatrix}0\\0\end{pmatrix} = 8$$

So the solution is that eight operations are required to achieve the target of 0.7 litres. The problem is a straightforward variation of the shortest path problem, with a series of nodes corresponding to the possible states of the two jugs ($(\binom{e}{f})$) that could be achieved and arcs of unit length (corresponding to one operation) linking them. One way of picturing the problem is shown in Figure 9.1 which uses a grid of equilateral triangles and a parallelogram of sides 5 and 8 to show the states. The only operations that are possible are those which follow grid lines from one edge of the figure to the other. The bold lines show the optimal procedure for obtaining 0.7 litres. By such a diagram, it is possible to picture some similar "transfer" problems, with other sizes of vessel, and there are analogues of it in three dimensions. So you could consider the problem of measuring exactly one litre with three jugs holding 0.3 litres, 0.6 litres and 0.8 litres, either as a dynamic programme or using a three dimensional grid.

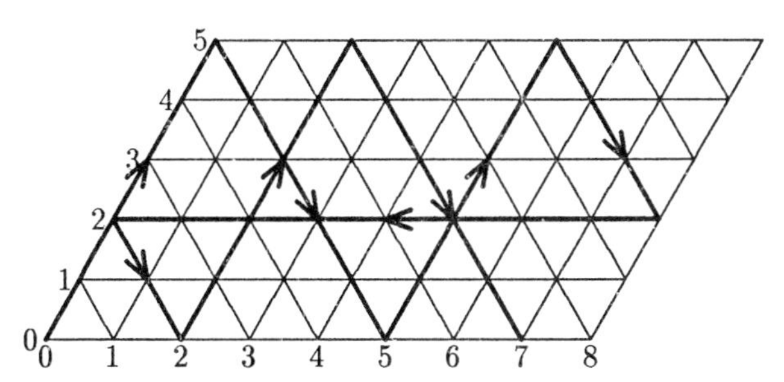

Figure 9.1: Graphical approach to problems of liquid measurement

Minimisation of the number of operations is not the only objective that might be used in this problem. We might want to minimise the amount of liquid that is poured away, or the amount of liquid that is taken from the source in order that the needs of the process be satisfied. The recurrence relations will be different in these cases, although the states and stages can remain the same.

9.3 River crossing

Another popular form of puzzle frequently appears in the form of arranging groups to cross a river. Usually it is posed as a story such as this:

> Three explorers are travelling in a jungle with three cannibals; they reach a wide river and build a raft that is big enough for two people to travel in. All six men want to cross the river, but if there are ever more cannibals than explorers on either bank, then the cannibals will rapidly consume the explorers. All the men can sail the raft equally well, and all will follow instructions exactly – cannibalism is the only blemish on the characters of anyone present. How should the party cross the river safely in the shortest possible time?

The relationship of such puzzles to dynamic programming is evident as one considers that there is a sequence of decisions to be made after each crossing; who should ride the raft for the next one? The objective, stages and states are very similar to those encountered in the pouring of liquids, but the transformation rules and constraints are different.

Suppose that we define the state of the system using the number of people (E_1, C_1) on the first bank of the river, with the empty raft there; there will be $(3-E_1, 3-C_1)$ on the opposite bank. A decision consists of choosing the people who will cross from the first bank to the second (e_1, c_1) followed by the choice (e_2, c_2) for the return journey. Clearly, there will only be a restricted set of values for the four numbers (e_1, c_1, e_2, c_2) since the raft cannot hold more than two people and the numbers of explorers must not be exceeded by the number of cannibals on either bank. The double crossing can be regarded as a stage in the sequential problem. We want to minimise the number of stages that are needed starting from the state $(3,3)$. So, as in the discussion about the pouring of liquid, we can define an objective function which represents the number of stages needed to achieve the target state $(0,0)$ under an optimal policy:

$$f^*(E_1, C_1) = 1 + \min\left[f^*(E_1 - e_1 + e_2, C_1 - c_1 + c_2)\right]$$

with the minimum taken over all the permissible sets of values of the four variables. There are only a limited number of feasible states for the problem, namely:

$$(0,0), (0,1), (0,2), (0,3), (1,1), (2,2), (3,1), (3,2), (3,3)$$

and the recurrence relationship yields:

$$\begin{aligned}
&f^*(0,0) = 0 \\
&f^*(0,1) = 1 \quad (c_1 = 1) \\
&f^*(0,2) = 1 \quad (c_2 = 2) \\
&f^*(0,3) = 1 + \min(f^*(2,2), f^*(0,2), f^*(1,1)) \\
&f^*(1,1) = 1 \quad (e_1 = c_1 = 1) \\
&f^*(2,2) = 1 + f^*(0,3) \quad (e_1 = 2, c_2 = 1) \\
&f^*(3,1) = 1 + \min(f^*(3,2), f^*(2,2)) \\
&f^*(3,2) = 1 + \min(f^*(3,1), f^*(3,3)) \\
&f^*(3,3) = 1 + f^*(3,2)
\end{aligned}$$

Solving these one has $f^*(3,3) = 6$, indicating that the raft must cross the river 11 times in all. An alternative formulation of this problem defines the states and stages in the same way, but defines the objective as $f_n^*(E_1, C_1)$ as the maximum number of people who can be left on the second bank using at most n stages from the given state. The interested reader may find this way of looking at the crossing problem on pages 97 to 99 of Bellman & Dreyfus (1962).

9.4 A gambling application

In several sections of this book, we have considered problems where the use of a stochastic term forced us to use the expected value of a variable as the objective. Stochastic models can have other objectives. One such occurs in some forms of gambling. The assumption is that the player has a limited number of opportunities (N) to gamble. (These may be caused by limited time, or be the result of the gambles being considered.) The gambler has both an initial sum of money ($£I$) and some target amount ($£T$) and wishes to try and maximise the probability that by end of the limited opportunities, that the target will have been reached. It is assumed that the probabilities of success are known and constant at each gambling opportunity. For simplicity, we shall suppose that there are two outcomes: to stake a sum $£x$ and with probability p to win and gain a reward of $£x$, or with probability $1-p$ to lose the stake money. Extensions to more complex situations are easy to develop.

In the dynamic programming formulation, it is easy to define the states and the stages. States will simply be the amount of money ($£s$) in hand; stages will be the number of opportunities (n) remaining. The objective, to be maximised, will be the probability of achieving the target before the opportunities run out. The decision to be made is the amount to stake in the next game, and an optimal policy will choose this and then follow an optimal scheme thereafter. The sums gambled must be whole pounds. Hence the recurrence relationship will take the form:

$$f_n^*(s) = \max_x \left[pf_{n-1}^*(s+x) + (1-p)f_{n-1}^*(s-x)\right]$$

with:

$$f_0^*(s) = 0(s < T) \qquad f_n^*(s) = 1(s \geq T)(n = 0, 1, \ldots, N)$$

As a trivial example, suppose the gamble is fair ($p = 0.5$) and the gambler's target is $£5$. What is the best strategy in five games, and how does the probability of achieving this vary with the amount with which the gambler starts?

Working backwards from $f_0^*()$, we can calculate $f_1^*()$ trivially. $f_1^*(5) = 1, f_1^*(4) = 0.5, f_1^*(3) = 0.5, f_1^*(2) = 0, f_1^*(1) = 0, f_1^*(0) = 0$. Going to the stage where there are two opportunities for gambling, a little more work is needed.

$$f_2^*(5) = 1$$
$$f_2^*(4) = \max\left[f_1^*(4), 0.5f_1^*(5) + 0.5f_1^*(3), 0.5f_1^*(6) + 0.5f_1^*(2)\right] = 0.75$$
$$f_2^*(3) = \max\left[f_1^*(3), 0.5f_1^*(4) + 0.5f_1^*(2), 0.5f_1^*(5) + 0.5f_1^*(1)\right] = 0.5$$
$$f_2^*(2) = \max\left[f_1^*(2), 0.5f_1^*(3) + 0.5f_1^*(1), 0.5f_1^*(4) + 0.5f_1^*(0)\right] = 0.25$$
$$f_2^*(1) = 0 \quad f_2^*(0) = 0$$

Values of $f_3^*()$ can be found in the same way, and then the dynamic programme works backwards again to give $f_4^*()$ and $f_5^*()$ Table 9.1 shows the values for all stages. It is a simple matter to identify the policy that yields these. For instance, if the gambler starts with $£3$, the first bet should be $£2$. Winning will finish the game; losing means that one has $£1$, and this must be gambled. If the outcome is a loss, that's the end of the game, otherwise the player has $£2$, and should gamble this, followed, if possible by a gamble of $£1$ – and (if necessary) a final bet of $£2$.

The probabilities in the final column ($f_5^*(s)$) are sensitive to the probability of success at each gamble. This follows from the way that the function values

Table 9.1: Probabilities of obtaining £5.00 in a fair gamble

state(s)	$f_0^*(s)$	$f_1^*(s)$	$f_2^*(s)$	$f_3^*(s)$	$f_4^*(s)$	$f_5^*(s)$
5	1.000	1.000	1.000	1.000	1.000	1.000
4	0.000	0.500	0.750	0.750	0.750	0.781
3	0.000	0.500	0.500	0.500	0.563	0.594
2	0.000	0.000	0.250	0.375	0.375	0.375
1	0.000	0.000	0.000	0.125	0.188	0.188
0	0.000	0.000	0.000	0.000	0.000	0.000

are calculated, to incorporate the possibility of several gambles with wins and losses.

9.5 Defective coins

A well-known puzzle, first published in the 19th century, asks for an optimal policy for weighing coins. One is given a bag of N coins which appear identical, but one is slightly different in weight from all the others. The coins can be weighed in a balance, one against one, two against two, etc.; what process of weighing will find the defective coin in the least number of balancing operations? Dynamic programming lends itself to the solution of such a problem because whenever a balancing operation has been completed, the set of coins containing the defective one will be reduced in size – and the set known to be good will be larger. So the problem is repeated (on a smaller scale) after each balancing operation. The decision to be taken at each stage is how many coins to select from each of these sets, and how they should be placed in the two pans of the balance. Bellman & Dreyfus (1962) carried out calculations for the problem where there are two defective coins; with more than two coins, the curse of dimensionality begins to make difficulties.

9.6 References to recreation in the dynamic programming literature

In many recreational activities, there is scope for using the ideas of dynamic programming. Sports and games frequently involve the participants in making sequential decisions over a period of time. These decisions can be considered within the general framework of a dynamic programme. Generally, sporting events involve randomness of some kind; a cricket stroke may or may not reach the boundary; a pass in a team game may or may not be intercepted; the next card to be turned up may or may not be the Ace of Diamonds; the next play of the gaming machine may or may not yield the "jackpot". There has been considerable research into potential applications of dynamic programming to the problems that are associated with strategy at sports. A serious difficulty is the scarcity of accurate data on the probabilities associated with the stochastic events that follow decisions. In this section we shall look at some of the ideas which research workers have developed and the problems that have been examined in applying dynamic programming to recreation and leisure activities.

Dynamic programming in one-day cricket – optimal scoring rates *Journal of the O.R. Society* **39** (1988) 4 (April) pp331-337 Clarke, S.R.

Stephen Clarke looked at problems of one-day or limited over cricket matches; here the number of opportunities for scoring runs is limited to a fixed number of balls (deliveries) and these form one possible description of the stages of a problem. The objectives of the two teams are well defined; the first team wants to score the maximum number of runs while not losing the wickets of all 10 of its players; the second team to play wants to score at least as many runs within its allotted number of deliveries. The main decision variable considered is the rate of scoring, because this determines both the expected rate of increase in the score and the risk of losing wickets. In many games, the conventional strategy is to score slowly at first and then try to increase the run rate towards the end of the innings; Clarke shows that, with his assumptions, this would not be optimal. As a dedicated follower of cricket, his analysis of his model is particularly impressive.

Dynamic programming in orienteering: route choice and the siting of controls *Journal of the O.R. Society* **35** (1984) 9 (September) pp791-796 Hayes, M. & Norman, J.M.

Orienteering has been described as "a sport for those who think" and this paper illustrates some of the decision processes that are involved in the choice of an optimal route from one checkpoint to the next. The authors were interested in several facets of the planning of orienteering events in the mountains of the Lake District in north-west England. They formulated the search for the best route as a path problem, with time allowances for travel across different kinds of ground and for ascending and descending slopes. Apart from the identification of the best routes, the authors pointed to the value of their model for the organisers of events; good and bad locations for checkpoints could be recognized, and so could suitable bases for first-aid teams assigned to deal with any injuries and accidents.

Dynamic policies in the long jump *Management Science in Sports* (1976) North-Holland, Amsterdam pp113-124 Sphicas, G.P. & Ladany, S.P.

and

Maximisation of the probability to score a given distance in a long jump competition *Computers and Operations Research* **14** (1987) 3 pp249- 256 Ladany, S.P. & Mehrez, A.

A long jumper aims to jump as far as possible beyond a marked take-off line on one of three attempts. A jump will only be counted if the actual take-off point was on – or in front of – the marked line. Therefore the jumper will aim to take off as close to the line as possible, without going beyond it. The authors assume that there is inherent inaccuracy in the long jumper's ability to judge a take-off point and in the length of the jump, and develop an optimal policy which depends on the stage (number of attempts taken so far) and the state (the maximum distance recorded). Ten years after this analysis appeared in the first of the papers cited above, Shaul Ladany published a follow-up analysis where the objective was to maximise the probability of scoring a given distance.

Optimal strategies for the game of darts *Journal of the O.R. Society* **33** (1982) 10 (October) pp871-884 Kohler, D.

David Kohler examined a popular version of the world-wide game of darts; players compete by throwing darts at a standard board marked into twenty segments, with certain areas marked for double or treble scores. The aim is to score 301 points before one's opponent, and to finish with a "Double". The author modelled the state of a player as the number of points remaining (301,300, ..., 4,3,2) and stages as a count of the number of darts thrown. A player's decisions are in two parts: which score to aim for, and where in the segment to aim for. Assuming that the player has some knowledge of his or her accuracy (and there is a guide to measuring that!) this paper presents guidelines on how to maximise your chances of winning at darts. (One thing is missing; instructions on how to consult this paper with ease in a smoke-filled bar when playing darts in the local club or public house.)

The basketball shoot-out: strategy and winning probabilities *Operations Research Letters* **5** (1986) 5 (November) pp241-244 Gerchak, Y. & Henig, M.

In a basketball shoot-out (as opposed to a match), players were asked to attempt one shot at the net from a distance of their own choice. The player who scored from the greatest distance was the winner. The order of players' shots was random, with later players knowing the current optimum score. The authors assumed that all competitors were equally skilled and showed that the probability of winning depended on the shooting order.

The inpolygon with minimal area and the circumpolygon with maximal perimeter *Optimization* **19** (1988) 2 pp229-233 Englisch, H. & Voigt, H.

This paper treats two (related) classical mathematical problems from a dynamic programming point of view. Given a polygon whose vertices are $(A_1, A_2, \ldots, A_n)$, the inpolygon is a polygon $(B_1, B_2, \ldots, B_n)$ whose vertices all satisfy the condition that B_i is on the edge A_iA_{i+1} of the original polygon. Two vertices of the inpolygon may coincide, and the angle subtended at a vertex may be a straight line. The original polygon is a circumpolygon for the inpolygon. The authors convert the geometrical problems into problems of determining optimal sequences of integers with specific properties. Dynamic programming provided the tool for finding such integer sequences.

Dynamic programming in tennis – when to use a fast serve *Journal of the O.R. Society* **36** (1985) 1 (January) pp75-77 Norman, J.M.

Several writers have considered the problem faced by a tennis player who can serve in two ways – "fast" and "slow". The former is more likely to defeat the opponent, but is more risky. Since the player with service has two chances to serve, it may be advantageous to serve "fast" first and then "slow". John Norman has analysed the problem as a dynamic programme and confirms the optimal strategy found by other writers using different approaches. The optimal strategy is independent of the state of the game, assuming that the probabilities remain constant. The structure of tennis scoring means that the dynamic programme could have an infinite number of stages, through repeated returns to "deuce".

An application of dynamic programming to pattern recognition *Journal of the O.R. Society* **37** (1986) 1 (January) pp35-40 Warwick, J. & Phelps, R.I.

Pattern recognition is one facet of the increasingly important science of artificial intelligence. The authors of this paper describe how a dynamic programming formulation was useful for helping to identify one kind of pattern, which has applications in the realm of biochemistry. This is a useful pointer to the way that the dynamic programming method can be used in fifth-generation computer software.

A dynamic programming formulation with diverse applications *Journal of the O.R. Society* **27** (1976) 1:i pp119-121 Adelson, R.M., Norman, J.M. & Laporte, G.

The authors were concerned with problems which required the permutation of the rows of a rectangular matrix in an optimal fashion. The matrix elements were all 0 or 1 and could be considered as "indicator variables" to show the absence or presence of some property. Row i of the matrix has an attached weight t_i. The objective of permuting the rows was to find the ordering of rows which minimised the total, over all the columns of the matrix, of the sum of the row weights starting at the first non-zero element in each column and ending at the last such element. Thus the matrix below, with weights given alongside, has an objective value 100:

$$\begin{matrix} 8 \\ 6 \\ 9 \\ 6 \end{matrix} \begin{pmatrix} 0 & 0 & 1 & 1 & 0 \\ 1 & 1 & 1 & 0 & 1 \\ 0 & 0 & 0 & 1 & 1 \\ 1 & 1 & 0 & 1 & 0 \end{pmatrix}$$

It can be permuted to give a smaller objective of 88:

$$\begin{matrix} 9 \\ 6 \\ 6 \\ 8 \end{matrix} \begin{pmatrix} 0 & 0 & 0 & 1 & 1 \\ 1 & 1 & 0 & 1 & 0 \\ 1 & 1 & 1 & 0 & 1 \\ 0 & 0 & 1 & 1 & 0 \end{pmatrix}$$

Such problems have applications in scheduling and in dating archaeological finds, amongst other areas. The authors devised a way of casting the permutation as a dynamic programme which proved more efficient than other approaches.

Optimal starting height for pole-vaulting *Operations Research* **23** (1975) pp968-978 Ladany, S.P.
and
Optimal pole-vaulting strategy *Operations Research* **37** (1989) 1 pp172-175 Hersh, M. & Ladany, S.P.

The rules of sports, unlike the laws of the Medes and Persians (Daniel chapter 6, verse 12), can be changed. When Shaul Ladany investigated a dynamic programming strategy for pole-vaulting in the 1970s, the competitor was required to make repeated attempts to vault the bar at a fixed height. The bar was raised by fixed increments and the principal decision faced by an athlete was the starting height for the bar. A decade later, these rules had changed, so that the conclusions reached in his 1975 paper were no longer valid. Competitors can now choose the starting height, and each successive setting of the bar, so long as these never decrease. Three failures in succession will eliminate a competitor, but these could be at different heights. An optimal strategy was devised, taking into account the effect of fatigue on a vaulter's prowess. The authors conclude with some observations on the effects of the changed rules – and of further, hypothetical ones, in case the rules of this sport are further amended!

Mini-Risk: strategies for a simplified board game *Journal of the O.R. Society* **41** (1990) 1 pp9-16 Maliphant, S.A. & Smith, D.K.

Sarah Maliphant examined some of the rules of the board game "Risk" as facets of a dynamic programme. The commercial product offers enormous scope for modelling decisions and strategy, and the published paper looks at some ways that lessons for modellers can be derived. One of the first problems to be answered was the definition of a suitable objective; when playing any board game, one wants to win. How is this to be translated into an objective for a single "round" of the game?

9.7 Exercises

(1) To settle a debt, a friend offers you a £10 note to pay an amount of £8.57. Formulate a dynamic programme which would enable you to calculate how to

give her change using the smallest number of coins/notes in general circulation in the U.K. (that is coins/notes valued at 1.00, 0.50, 0.20, 0.10, 0.05, 0.02, 0.01). How would this solution be affected if you were using American coins/notes (1.00, 0.25, 0.10, 0.05, 0.01)? Or Swedish (1.00, 0.50, 0.10, 0.05, 0.01)?

(2)The game of Nim is played between two people, with a number of piles of sticks on the table between them. The players take turns to remove some or all of the sticks from one of the piles. The player who picks up the last stick wins the game. Thus a game might develop as follows:

IIIIIII III I (player 1 removes 2 sticks from the second pile)
IIIIIII I I (player 2 removes 5 sticks from the first pile)
II I I (player 1 removes all the first pile)
I I (player 2 removes all the second pile)
I (player 1 wins)

How can dynamic programming be used to determine the optimal policy for a player?

(3) The game of "Bulls and Cows" (also known by a variety of trade names) takes several forms. The rules are essentially the same, whatever variant is being played:

1: Player A places one item from a set of M different designs in each of n slots, which are ordered $1 \ldots n$ to form a "pattern". The items may be pegs, characters or digits. The slots are hidden from Player B.
2: Player B tries to guess the "pattern" that Player A has created and forms a "pattern".
3: Player A examines Player B's "pattern" and informs Player B of the total number of the following:
   α: Items which match exactly; these are the correct design and in the correct slots; this number is referred to as the number of "Bulls".
   β: Items which match, but not exactly; they are of the correct design but in the wrong slot; this total is the number of "Cows".
4: Play continues, repeating steps 2 and 3 until Player B matches the pattern exactly, by getting a score of n "Bulls".
5: Player B has an objective of trying to minimise the number of patterns that need to be compared before the game finishes.

Suppose that $M = 3$ and $n = 2$; formulate a dynamic programme for Player B's strategy. Explore how it could be extended to larger games. Explore what strategy Player A should adopt in order to make the game as hard as possible for the second player, where you define "hard" in some sense which could be measured.

(4) Two players take turns to roll dice. The first player rolls three dice and records the largest two scores, A_1 and A_2, with $A_1 \geq A_2$. The second player decides to roll either one or two dice and scores B_1 and possibly B_2 with $B_1 \geq B_2$.

If $A_1 > B_1$ the second player pays the first one £1.00, otherwise the first player pays the second player the same amount. The same rules apply, if necessary, to A_2 and B_2. Play continues until the first player has £2.00 or the second has £1.00 (indicating that it would be impossible to continue) and then the player with the larger amount of money wins. What decision rules should the second player adopt in order to maximise his chance of winning?†

(5) Four explorers and four cannibals want to cross a river where there is a raft which can hold three people. As in the problem discussed, the number of cannibals on either bank of the river must never exceed the number of explorers, nor may the crew of the raft be 2 cannibals and 1 explorer. Formulate the problem of crossing as a dynamic programme, and find the sequence of passengers which minimises the number of times that the raft crosses the river.

(6) You have £3 in your pocket, and you want to try and double this by a gamble with three outcomes: for a stake of £x you gain £x, lose £x or gain nothing, each with probability $\frac{1}{3}$; what is the best strategy for you to follow if you can play at most five games?

(7) Formulate and solve the problem of identifying one defective coin from a bag of 8, using dynamic programming to minimise the expected number of balancing operations.

(8) "If you don't do your best with what you've got, you will never do better later than you would have done if you had done your best now." Do you think that this statement is an accurate summing-up of the ideas of dynamic programming?

† I am indebted to Christopher Dearlove for this idea, which arose from his analysis of a variant of the rules of "Risk".

Appendix

Computers and Dynamic Programming

Black boxes?

From the outset, this book has stressed that dynamic programming is not a problem-solving technique which can be created as a "Black box" – numbers input and solutions output, without much thought in between. The emphasis in all the chapters has been on the formulation of problems, leading to a recurrence relationship which can then be used for hand or automatic calculation. The exercises at the ends of the chapters were designed to be solved by hand, which may have made them appear somewhat unrealistic in size; similarly, the worked examples, for the most part, were solved by hand. The variety of the contents of the book should indicate the range of different formulations that are possible.

Nonetheless, there have been a number of attempts to develop computer programs to help both the student and the professional using dynamic programming. In this appendix, brief notes are given regarding some of these.

Computer programs from books

Optimization techniques with Fortran (1973) Mc-Graw Hill (New York) 0-07-035606-8 Kuester, James L. & Mize, Joe H.

Although this book has long been out of print, it deserves to be mentioned for the great scope of its computer codes. Copies are likely to be found in many OR reference libraries. In the 1990s, programs designed to be run as batch jobs (from cards) using Fortran IV may seem as relevant as programs written in Latin; but, with modifications to suit an era of interactive programming, these programs are still valuable. The authors cover many other optimisation techniques than dynamic programming, but devote about fifty pages to listing a suite of programs for DP. These are divided into two parts corresponding to discrete and continuous variables.

For the discrete variables, nine recurrence relationships are offered:

$$f_n^*(s) = \min_x \left(C_{sx} + f_{n-1}^*(x)\right) \tag{10.01}$$

$$f_n^*(s) = \max_x \left(P_{nx} + f_{n-1}^*(s-x)\right) \tag{10.02}$$

$$f_n^*(s) = \max_x \left(P_{nx} \times f_{n-1}^*(s-x)\right) \tag{10.03}$$

$$f_n^*(s) = \max_x \left(p \times f_{n-1}^*(s-x) + (1-p) \times f_{n-1}^*(s+x)\right) \tag{10.04}$$

$$f_n^*(s) = \min_x \left(\phi(x) + \sum_{k=0}^{k=K} \left(\binom{x}{k} p^k (1-p)^{K-k} f_{n-1}^*(s-k)\right)\right) \tag{10.05}$$

$$f_n^*(s) = \min_x \left(C_1(x-s)^2 + C_2(x-M_n) + f_{n-1}^*(x)\right) \tag{10.06}$$

$$f_n^*(s) = \min_x \left(xP_n + f_{n-1}^*(s-xM_n)\right) \tag{10.07}$$

$$f_n^*(s) = \min_x \left(P(x) + F(x) + hs + f_{n-1}^*(s+x-D_n)\right) \tag{10.08}$$

$$f_n^*(s) = \max_x \left(P(n,x) \times f_{n-1}^*(s-M_n(x))\right) \tag{10.09}$$

Most of these have appeared in one form or another in this book.

(10.01) is simply the problem of finding a shortest path, with C_{sx} the cost of travel from node s to node x. Chapter 2 has the details.

(10.02) is a knapsack problem, with P_{nx} the benefit of assigning x units of resource to the nth item on the list. (Chapter 3).

(10.03) is a multiplicative knapsack problem where P_{nx} is the probability of success of the nth stage if x units of resource are allocated to it. The objective is to maximise the probability of the success of the project (exercise 10 in chapter 3 is an instance of this).

(10.04) is the recurrence relation for the coin-tossing game described in chapter 9.

The authors present equation (10.05) as a production problem, where p is the probability of a good item being made; s is the number of items still needed at stage n, and the upper limit of the summation K is defined to be the smaller of x and s; thus, it is a variation of the problems presented in section 6.5.

(10.06) has not appeared in this book; it was devised as a method for deciding when to hire or fire temporary staff in employment where the numbers required are volatile. In time period n, a minimum of M_n staff will be needed. If s are available at the start of the period, and a decision is made to employ x during the period, then there will be two costs incurred. The first is quadratic, relating to the cost of either hiring $(x - s)$ or firing $(s - x)$ workers; the second is the cost of employing the surplus workers for the nth period.

Equation (10.07) appears in the book by Hillier & Lieberman (1967) as an application of dynamic programming to the solution of linear programming problems.

The pattern of equation (10.08) is that of the production scheduling problem encountered in chapter 4, with the costs of producing at a given stage and holding costs (but these are based on the opening inventory, not the closing one).

Finally, equation (10.09) is an extension of (10.03) where s measures the money available for a project, and not simply the total resources available. $M_n(x)$ is the cost of using x units of resource n.

The authors base these formulations on the presentation of dynamic programming given by Hillier & Lieberman (1967) and on the work of the research student Rider (1971). They admit that there are many other formulations possible, as has been shown in the chapters of this book. The second part of their presentation of computer solutions is concerned with the optimisation of nonlinear functions in several dimensions, covered in chapter 5.

There are elementary worked examples of both the continuous and the discrete problems.

Computational techniques in operations research (1985) Abacus Press (Tunbridge Wells) 0-85626-425-3 Andrew, A.M.

One chapter of this slim book is devoted to dynamic programming. After a brief discussion of the concepts, attention is shifted to Howard's method. The author lists a program in PASCAL for solving problems of Markov processes with gains. Should this type of problem need to be solved regularly, then the reader may find it helpful to have access to a copy of this text. There is a detailed review of the whole book in *Journal of the O.R. Society* **37** (1986) 6 (June) p650-651

On a language for discrete dynamic programming and a microcomputer implementation *Computers and Operations Research* **16** (1989) 1 p1-11 Deuermeyer, Bryan L.

This is one of the most radical ideas in recent years for those who regularly use dynamic programming. The author has developed both a computer language and a computing environment to help to solve discrete dynamic programming problems. In the article, he presents both the strategy behind this research and some instances of its use to some of the problems that have been examined in this book. The user has to consider one decision problem, which could be viewed as a stochastic decision problem: is it worth expending the effort to learn a new set of programming rules in order to be able to solve dynamic programmes more easily? The uncertainty involved in the decision process concerns the potential savings of time that such specialised knowledge would bring. The reasoning behind the development of the structures of the language are useful, even for those who choose to develop programs in other computer languages.

The Exeter suite of programs

While writing this book, the author developed a suite of Pascal programs to run under Turbo Pascal on an IBM personal computer. These cover many of the formulations that have been covered, and have been designed to enable both rapid solution of problems and variation of the parameters in a convenient form. The author is happy for these programs to be used by the readers of the book; they are development tools, not commercial packages, and are available on this understanding. To run them, you will need Turbo Pascal (version 4.0 or higher) on an IBM PC or equivalent. Copies of the programs can be obtained from the author by sending an electronic mail message to `dks@msor.exeter.ac.uk` in which case the suite will be sent by electronic mail, or by sending a blank diskette to the author with a suitable sum for return postage.

The programs cover the models from the following sections of the book: 2.1, 3.2 and 3.5, 4.4, 4.7, 6.5, 7.2, 7.5, 8.4, 9.4. No warranty of any kind is implied with the programs, but they are made available in good faith for those who want to extend their knowledge of dynamic programming.

Bibliography and Further Reading

General reading

The literature of dynamic programming is a vast one; in these pages I have chosen to point to some of the literature which has been of most help to me, in my studies. The books that are mentioned below are those which cover the range of dynamic programming problems best; all are suitable for students at undergraduate level or higher.

(1) Bellman, Richard E. (1957) *Dynamic programming,* Princeton U.P.
(2) Bellman, Richard E. & Dreyfus, Stuart E. (1962) *Applied dynamic programming,* Princeton U.P. and Oxford U.P.
(3) Hastings, Nicholas A.J. (1973) *Dynamic programming with management applications,* Butterworths
(4) Larson, Robert E. & Casti, John L. (1978) & (1982) *Principles of dynamic programming, volume 1 and volume 2* Dekker
(5) Nemhauser, G.L. (1966) *Introduction to dynamic programming,* John Wiley
(6) Norman, John M. (1972) *Heuristic procedures in dynamic programming,* Manchester University Press
(7) Norman, John M. (1975) *Elementary dynamic programming,* Edward Arnold
(8) Whittle, Peter (1982) *Optimization over time: dynamic programming and stochastic control, (Volumes I and II),* John Wiley

Specific references

The books and articles below have been cited in the body of the text: this list does not duplicate the entries in section 9.6 and those in the Appendix. In addition to the papers that are quoted here, there is a steady stream of journal articles illustrating new applications and developments in dynamic programming. The International Abstracts in Operations Research and the OR/MS index will be good places to start looking for the most recent progress, whether theoretical or practical, in dynamic programming.

(9) Black, W.L. (1965) Discrete sequential search, *Information and Control.* **8.** p159–162

(10) French, Simon (1981) *Sequencing and scheduling: an introduction to the mathematics of the job shop*, Ellis Horwood

(11) Graham, Ronald L., Knuth, Donald E. & Patashnik, Oren (1989) *Concrete mathematics: a foundation for computer science,* Addison-Wesley

(12) Hillier, Frederick S. & Lieberman, Gerald J. (1967) *Introduction to operations research,* Holden-Day

(13) Howard, Ronald A. (1960) *Dynamic programming and Markov processes,* MIT Press & John Wiley

(14) Johnson, Lynwood A. & Montgomery, Douglas C. (1974) *Operations research in production planning, scheduling, and inventory control,* Wiley

(15) Karush, W. (1939) *Minima of functions of several variables with inequalities as side conditions,* M.S. Thesis, Dept of Mathematics, University of Chicago.

(16) Knuth, Donald E. (1984) *The TEXbook,* Addison-Wesley

(17) Knuth, Donald E. (1969) (1981) & (1973) *The art of computer programming, Volume 1: Fundamental algorithms, Volume 2: Seminumerical algorithms (2nd edition), Volume 3: Sorting and searching,* Addison-Wesley

(18) Kuhn, W.W. & Tucker, A.W. (1951) Nonlinear Programming. In *Proceedings of the 2nd Berkeley Symposium on Mathematical Statistics and Probability,* University of California Press, pp481-492

(19) Rider, E. (1971) *General computer solution of dynamic programming problems with integer restrictions,* M.S. Thesis, Arizona State University

(20) Vajda, Steven (1989) *Fibonacci and Lucas numbers, and the golden section: theory and applications,* Ellis Horwood

Index

Mathematics and its Applications

Series Editor: G. M. BELL,

Professor of Mathematics, King's College London, University of London

Ball, M.A.	**Mathematics in the Social and Life Sciences: Theories, Models and Methods**
de Barra, G.	**Measure Theory and Integration**
Bartak, J., Herrmann, L., Lovicar, V. & Vejvoda, D.	**Partial Differential Equations of Evolution**
Bejancu, A.	**Finsler Geometry and Applications**
Bell, G.M. and Lavis, D.A.	**Statistical Mechanics of Lattice Models, Vols. 1 & 2**
Berry, J.S., Burghes, D.N., Huntley, I.D., James, D.J.G. & Moscardini, A.O.	**Mathematical Modelling Courses**
Berry, J.S., Burghes, D.N., Huntley, I.D., James, D.J.G. & Moscardini, A.O.	**Mathematical Modelling Methodology, Models and Micros**
Berry, J.S., Burghes, D.N., Huntley, I.D., James, D.J.G. & Moscardini, A.O.	**Teaching and Applying Mathematical Modelling**
Blum, W.	**Applications and Modelling in Learning and Teaching Mathematics**
Brown, R.	**Topology: A Geometric Account of General Topology, Homotopy Types and the Fundamental Groupoid**
Bunday, B. D.	**Statistical Methods in Reliability Theory and Practice**
Burghes, D.N. & Borrie, M.	**Modelling with Differential Equations**
Burghes, D.N. & Downs, A.M.	**Modern Introduction to Classical Mechanics and Control**
Burghes, D.N. & Graham, A.	**Introduction to Control Theory, including Optimal Control**
Burghes, D.N., Huntley, I. & McDonald, J.	**Applying Mathematics**
Burghes, D.N. & Wood, A.D.	**Mathematical Models in the Social, Management and Life Sciences**
Butkovskiy, A.G.	**Green's Functions and Transfer Functions Handbook**
Cartwright, M.	**Fourier Methods: for Mathematicians, Scientists and Engineers**
Cerny, I.	**Complex Domain Analysis**
Chorlton, F.	**Textbook of Dynamics, 2nd Edition**
Chorlton, F.	**Vector and Tensor Methods**
Cohen, D.E.	**Computability and Logic**
Cordier, J.-M. & Porter, T.	**Shape Theory: Categorical Methods of Approximation**
Crapper, G.D.	**Introduction to Water Waves**
Cross, M. & Moscardini, A.O.	**Learning the Art of Mathematical Modelling**
Cullen, M.R.	**Linear Models in Biology**
Dunning-Davies, J.	**Mathematical Methods for Mathematicians, Physical Scientists and Engineers**
Eason, G., Coles, C.W. & Gettinby, G.	**Mathematics and Statistics for the Biosciences**
El Jai, A. & Pritchard, A.J.	**Sensors and Controls in the Analysis of Distributed Systems**
Exton, H.	**Multiple Hypergeometric Functions and Applications**
Exton, H.	**Handbook of Hypergeometric Integrals**
Exton, H.	***q*-Hypergeometric Functions and Applications**
Faux, I.D. & Pratt, M.J.	**Computational Geometry for Design and Manufacture**
Firby, P.A. & Gardiner, C.F.	**Surface Topology**
Gardiner, C.F.	**Modern Algebra**
Gardiner, C.F.	**Algebraic Structures**
Gasson, P.C.	**Geometry of Spatial Forms**
Goodbody, A.M.	**Cartesian Tensors**
Graham, A.	**Kronecker Products and Matrix Calculus: with Applications**
Graham, A.	**Matrix Theory and Applications for Engineers and Mathematicians**
Graham, A.	**Nonnegative Matrices and Applicable Topics in Linear Algebra**
Griffel, D.H.	**Applied Functional Analysis**
Griffel, D.H.	**Linear Algebra and its Applications: Vol. 1, A First Course; Vol. 2, More Advanced**
Guest, P. B.	**The Laplace Transform and Applications**
Hanyga, A.	**Mathematical Theory of Non-linear Elasticity**
Harris, D.J.	**Mathematics for Business, Management and Economics**
Hart, D. & Croft, A.	**Modelling with Projectiles**
Hoskins, R.F.	**Generalised Functions**
Hoskins, R.F.	**Standard and Nonstandard Analysis**
Hunter, S.C.	**Mechanics of Continuous Media, 2nd (Revised) Edition**
Huntley, I. & Johnson, R.M.	**Linear and Nonlinear Differential Equations**
Irons, B. M. & Shrive, N. G.	**Numerical Methods in Engineering and Applied Science**
Ivanov, L. L.	**Algebraic Recursion Theory**
Johnson, R.M.	**Theory and Applications of Linear Differential and Difference Equations**
Johnson, R.M.	**Calculus: Theory and Applications in Technology and the Physical and Life Sciences**
Jones, R.H. & Steele, N.C.	**Mathematics in Communication Theory**
Jordan, D.	**Geometric Topology**
Kelly, J.C.	**Abstract Algebra**
Kim, K.H. & Roush, F.W.	**Applied Abstract Algebra**
Kim, K.H. & Roush, F.W.	**Team Theory**
Kosinski, W.	**Field Singularities and Wave Analysis in Continuum Mechanics**
Krishnamurthy, V.	**Combinatorics: Theory and Applications**
Lindfield, G. & Penny, J.E.T.	**Microcomputers in Numerical Analysis**
Livesley, K.	**Mathematical Methods for Engineers**
Lord, E.A. & Wilson, C.B.	**The Mathematical Description of Shape and Form**
Malik, M., Riznichenko, G.Y. & Rubin, A.B.	**Biological Electron Transport Processes and their Computer Simulation**
Martin, D.	**Manifold Theory: An Introduction for Mathematical Physicists**
Massey, B.S.	**Measures in Science and Engineering**
Meek, B.L. & Fairthorne, S.	**Using Computers**

Mathematics and its Applications

Series Editor: G. M. BELL,
Professor of Mathematics, King's College London, University of London

- Menell, A. & Bazin, M. **Mathematics for the Biosciences**
- Mikolas, M. **Real Functions and Orthogonal Series**
- Moore, R. **Computational Functional Analysis**
- Moshier, S.L.B. **Methods and Programs for Mathematical Functions**
- Murphy, J.A., Ridout, D. & McShane, B. **Numerical Analysis, Algorithms and Computation**
- Nonweiler, T.R.F. **Computational Mathematics: An Introduction to Numerical Approximation**
- Ogden, R.W. **Non-linear Elastic Deformations**
- Oldknow, A. **Microcomputers in Geometry**
- Oldknow, A. & Smith, D. **Learning Mathematics with Micros**
- O'Neill, M.E. & Chorlton, F. **Ideal and Incompressible Fluid Dynamics**
- O'Neill, M.E. & Chorlton, F. **Viscous and Compressible Fluid Dynamics**
- Page, S. G. **Mathematics: A Second Start**
- Prior, D. & Moscardini, A.O. **Model Formulation Analysis**
- Rankin, R.A. **Modular Forms**
- Scorer, R.S. **Environmental Aerodynamics**
- Shivamoggi, B.K. **Stability of Parallel Gas Flows**
- Srivastava, H.M. & Manocha, L. **A Treatise on Generating Functions**
- Stirling, D.S.G. **Mathematical Analysis**
- Sweet, M.V. **Algebra, Geometry and Trigonometry in Science, Engineering and Mathematics**
- Temperley, H.N.V. **Graph Theory and Applications**
- Temperley, H.N.V. **Liquids and Their Properties**
- Thom, R. **Mathematical Models of Morphogenesis**
- Toth, G. **Harmonic and Minimal Maps and Applications in Geometry and Physics**
- Townend, M. S. **Mathematics in Sport**
- Townend, M.S. & Pountney, D.C. **Computer-aided Engineering Mathematics**
- Trinajstic, N. **Mathematical and Computational Concepts in Chemistry**
- Twizell, E.H. **Computational Methods for Partial Differential Equations**
- Twizell, E.H. **Numerical Methods, with Applications in the Biomedical Sciences**
- Vince, A. and Morris, C. **Discrete Mathematics for Computing**
- Walton, K., Marshall, J., Gorecki, H. & Korytowski, A. **Control Theory for Time Delay Systems**
- Warren, M.D. **Flow Modelling in Industrial Processes**
- Wheeler, R.F. **Rethinking Mathematical Concepts**
- Willmore, T.J. **Total Curvature in Riemannian Geometry**
- Willmore, T.J. & Hitchin, N. **Global Riemannian Geometry**

Statistics, Operational Research and Computational Mathematics

Editor: B. W. CONOLLY,
Emeritus Professor of Mathematics (Operational Research), Queen Mary College, University of London

- Abaffy, J. & Spedicato, E. **ABS Projection Algorithms: Mathematical Techniques for Linear and Nonlinear Equations**
- Beaumont, G.P. **Introductory Applied Probability**
- Beaumont, G.P. **Probability and Random Variables**
- Conolly, B.W. **Techniques in Operational Research: Vol. 1, Queueing Systems**
- Conolly, B.W. **Techniques in Operational Research: Vol. 2, Models, Search, Randomization**
- Conolly, B.W. **Lecture Notes in Queueing Systems**
- Conolly, B.W. & Pierce, J.G. **Information Mechanics: Transformation of Information in Management, Command, Control and Communication**
- French, S. **Sequencing and Scheduling: Mathematics of the Job Shop**
- French, S. **Decision Theory: An Introduction to the Mathematics of Rationality**
- Goult, R.J. **Applied Linear Algebra**
- Griffiths, P. & Hill, I.D. **Applied Statistics Algorithms**
- Hartley, R. **Linear and Non-linear Programming**
- Jolliffe, F.R. **Survey Design and Analysis**
- Jones, A.J. **Game Theory**
- Kapadia, R. & Andersson, G. **Statistics Explained: Basic Concepts and Methods**
- Lootsma, F. **Operational Research in Long Term Planning**
- Moscardini, A.O. & Robson, E.H. **Mathematical Modelling for Information Technology**
- Moshier, S.L.B. **Mathematical Functions for Computers**
- Norcliffe, A. & Slater, G. **Mathematics of Software Construction**
- Oliveira-Pinto, F. **Simulation Concepts in Mathematical Modelling**
- Ratschek, J. & Rokne, J. **New Computer Methods for Global Optimization**
- Schendel, U. **Introduction to Numerical Methods for Parallel Computers**
- Schendel, U. **Sparse Matrices**
- Sehmi, N.S. **Large Order Structural Eigenanalysis Techniques: Algorithms for Finite Element Systems**
- Sewell, G. **Computational Methods of Linear Algebra**
- Sharma, O.P. **Markovian Queues**
- Smith, D.K. **Dynamic Programming: A Practical Introduction**
- Späth, H. **Mathematical Software for Linear Regression**
- Stoodley, K.D.C. **Applied and Computational Statistics: A First Course**
- Stoodley, K.D.C., Lewis, T. & Stainton, C.L.S. **Applied Statistical Techniques**
- Thomas, L.C. **Games, Theory and Applications**
- Vajda, S. **Fibonacci and Lucas Numbers, and the Golden Section**
- Whitehead, J.R. **The Design and Analysis of Sequential Clinical Trials**